The Revenge of Gaia

JAMES LOVELOCK

The Revenge of Gaia

Earth's Climate in Crisis and the Fate of Humanity

Foreword by Sir Crispin Tickell

A Member of the Perseus Books Group
New York

Published by Basic Books,
A Member of the Perseus Books Group
First published in the United Kingdom in 2006 by Allen Lane,
A Member of the Penguin Group.

A CIP catalog record for this book is available from the Library of Congress
ISBN-13: 978-0-465-04168-8
ISBN-10: 0-465-04168-X
British ISBNs: 978-0-713-99914-3; 0-713-99914-4
06 07 08 09 / 10 9 8 7 6 5 4 3 2 1

I dedicate this book to my beloved wife Sandy

Contents

List of Illustrations

(Photographic acknowledgements are given in parentheses.)

Acknowledgements

I have been fortunate to have friends who read and who made helpful and valued comments on the book as it was written and I am truly grateful to: Richard Betts, David Clemmow, Peter Cox, John Dyson, John Gray, Stephan Harding, Peter and Jane Horton, Tim Lenton, Peter Liss, Chris Rapley, John Ritch, Elaine Steel, Sir Crispin Tickell, David Ward and Dave Wilkinson. I also thank GAIA, registered charity no. 327903, www.daisyworld.org, for support during the writing of this book and to whom all royalties will be donated.

Preface to the U.S. Edition

One of the hardest tasks we face in life is to be the bearer of seriously bad news. No one knows this more than the army officer tasked to tell a family that their son or daughter had died in action. This has been the hardest of books to write for the same reason. I have for the past forty years looked on the Earth through Gaia theory as if, metaphorically, it were alive at least in the sense that it regulates climate and composition of the Earth's surface so as always to be fit for whatever forms of life inhabit it. It is not pushing the metaphor too far to consider anything alive as either healthy or diseased. Thinking this way has made me a member of the new profession of planetary physicians, and as a planetary doctor I have now to bring the worst of news. The climate centers around the world, which are the equivalent of pathology labs in hospitals, have reported the Earth's physical condition, and the climate specialists see it as seriously ill and soon to pass into a morbid fever that may last as long as 100,000 years. I have to tell you, as intimate members of the Earth's family, that civilization is in grave danger.

Without our realising it we have poisoned the earth by our emissions of greenhouse gases and weakened it by taking for farmland and housing the land that once was the home of ecosystems that sustained the environment. We have driven the Earth to a crisis state from which it may never, on a human time scale, return to the lush and comfortable world we love and in which we grew up.

This is no sci-fi speculation; we now have evidence from the Earth's history that a similar event happened fifty-five million years ago when a geological accident released into the air more than a terraton of gaseous carbon compounds. As a consequence the temperature in the arctic and temperate regions rose eight degree Celsius and in tropical regions about five degrees, and it took over one hundred thousand years before normality was restored. We have already put more than half this quantity of carbon gases into the air and now the Earth is weakened by the loss of land we took to feed and house ourselves. In addition, the sun is now warmer, and as a consequence the Earth is now returning to the hot state it was in before, millions of years ago, and as it warms, most living things will die. Once started, the move to a hot state is irreversible, and even if all the good intentions expressed at the Kyoto and Montreal meetings were executed immediately, they would not alter the outcome. Much of the tropical land mass will become scrub and desert and will no longer serve for regulation, thus adding to the 40 percent of the Earth's surface we have already depleted to feed ourselves. Curiously, smoke and dust pollution of the northern hemisphere reduces global warming by reflecting sunlight back to space. This 'global dimming' is transient and could disappear in a few days if there were an economic downturn or a reduction of fossil fuel burning. This would leave us fully exposed to the heat of the global greenhouse. We are in a fool's climate, accidentally kept cool by smoke, and before this century is over, billions of us will die and the few breeding pairs of people that survive will be in the arctic region where the climate remains tolerable.

The great party of the twentieth century is coming to an end, and unless we now start preparing our survival kit we will soon be just another species eking out an existence in the few remaining habitable regions. Perhaps the saddest thing is that Gaia will lose as much or more than we do. Not only will wildlife and whole ecosystems go extinct, but the planet will lose a precious resource: human civilization. Humans are not merely a disease; we are, through our intelligence and communication, the nervous system of the planet. Through us Gaia has seen herself from space and begins to know her place in the universe. We should be the heart and mind of the Earth, not its malady. So let us be brave and cease thinking of human needs and rights alone and see that we have harmed the living Earth and need to make our peace with Gaia. We must do it while we are still strong enough to negotiate and not a broken rabble led by brutal war lords. Most of all we should remember that we are a part of Gaia, and she is indeed our home.

Foreword

Who is Gaia? What is she? The What is the thin spherical shell of land and water between the incandescent interior of the Earth and the upper atmosphere surrounding it. The Who is the interacting tissue of living organisms which over four billion years has come to inhabit it. The combination of the What and the Who, and the way in which each continuously affects the other, has been well named 'Gaia'. It is, as James Lovelock says, a metaphor for the living Earth. The Greek goddess from whom the term is derived should be proud of the use to which her name has been put.

The notion that the Earth is in this metaphorical sense alive has a long history. Gods and goddesses were seen to embody specific elements, ranging from the sky to the most local spring, and the notion that the Earth itself was alive came up regularly in Greek philosophy. Leonardo da Vinci saw the human body as the microcosm of the Earth, and the Earth as the macrocosm of the human body. He did not know as well as we now do that the human body is a macrocosm of the tiny elements of life – bacteria, parasites, viruses – often at war with each other, and together constituting more than our body cells. Giordano Bruno was burnt at the stake just over 400 years ago for maintaining that the Earth was alive, and that other planets could be so too. The geologist James Hutton saw the Earth as a self-regulating system in 1785, and T. H. Huxley saw it likewise in 1877. For his part, Vladimir Ivanovich Vernadsky saw the functioning of the biosphere as a geological force which creates a dynamic disequilibrium which in turn promotes the diversity of life.

But it was James Lovelock who brought this together into the Gaia hypothesis in 1972. In this book he refines and enlarges upon it in new and practical ways. Looking back it is strange how uncongenial the idea was to the practitioners of the conventional wisdom when it was put forward in its present form over a quarter of a century ago. Unfamiliar ways of looking at the familiar tend to arouse emotional opposition far beyond rational argument: thus the opposition to the ideas of evolution by natural selection in the nineteenth century, of tectonic plate movement in the twentieth century, and more recently of Gaia. At the beginning some New Age travellers jumped aboard, and some otherwise sensible scientists jumped off. They are now jumping on again. The change was well summed up in a declaration published after a meeting of scientists from the four great international global research programmes in 2001 which said

> The Earth system behaves as a single, self-regulating system, comprised of physical, chemical, biological and human components. The interactions and feedbacks between the component parts are complex and exhibit multi-scale temporal and spatial variability.

This indeed is Gaia.

The prime message from this book is less that Gaia herself is under threat ('a tough bitch', as Lynn Margulis has called her), but rather that humans have been doing her present configuration increasingly serious damage. Gaia is anyway changing, and may be less robust than in the past. The sun's heat on the Earth is steadily increasing, and eventually the self-regulation on which all life depends will be put at risk. Looking at the global ecosystem as a whole, human population increase, degradation of land, depletion of resources, accumulation of wastes, pollution of all kinds, climate change, abuses of technology, and destruction to biodiversity in all its forms together constitute a unique threat to human welfare unknown to previous generations. As Lovelock has written elsewhere,

> We have grown in number to the point where our presence is perceptibly disabling the planet like a disease. As in human diseases there are four possible outcomes: destruction of the invading disease organisms; chronic

infection; destruction of the host; or symbiosis – a lasting relationship of mutual benefit to the host and the invader.

The question is how to achieve that symbiosis. We are far from it today. Lovelock eloquently examines each of the main issues, most arising from the effects of the industrial revolution, in particular use of fossil fuels, chemicals, agriculture and living space. He then goes on to suggest how we might – at long last – begin to cope. As has been well said, the first requirement is to recognize that the problems exist. The second is to understand and draw the right conclusions. The third is to do something about them. Today we are somewhere between stages one and two.

When applied to the problems of present society, the concept of Gaia can be extended to current thinking about values: the way we look at and judge the world around us, and above all how we behave. This has particular application in the field of economics, where fashionable delusions about the supremacy of market forces are so deeply entrenched, and the responsibility of government to protect the public interest is so often ignored. Rarely do we measure costs correctly: thus the mess of current energy and transport policy, and the failure to assess the likely impacts of climate change.

The main difference between the past and today is that our problems are truly global. As Lovelock points out, we are currently trapped in a vicious circle of positive feedback. What happens in one place very soon affects what happens in others. We are dangerously ignorant of our own ignorance, and rarely try to see things as a whole. If we are eventually to achieve a human society in harmony with nature, we must be guided by more respect for it. No wonder that some have wanted to make a religion of Gaia, or of life as such. This book is a marvellous introduction to the science of how our species should make its peace with the rest of the world in which we live.

CRISPIN TICKELL

NOTE

The symbol † indicates that further definition is given in the glossary (pp. 160–65).

I

The State of the Earth

Ye blind guides, which strain at a gnat, and swallow a camel.
King James Bible, Matthew 23:24

As always, bad events usurp the news agenda, and as I write in the comfort of my Devon home, the New Orleans catastrophe fills the television screens and front pages. Horrific though it was, it distracts us from the more extensive suffering caused by the tsunami in December 2004 that disastrously splashed across the bowl of the Indian Ocean. That awful event starkly revealed the power of the Earth to kill. The planet we live on has merely to shrug to take some fraction of a million people to their death. But this is nothing compared with what may soon happen; we are now so abusing the Earth that it may rise and move back to the hot state it was in fifty-five million years ago, and if it does most of us, and our descendants, will die. It is as if we were committed to live through the mythical tale of Wagner's *Der Ring des Nibelungen* and see our Valhalla melt in torrid heat.

But I hear you say, 'What? Another book on global warming; isn't what was once a scare now becoming overkill?' If this book were no more than a reiteration of the arguments and counterarguments you would be right, and it would be one book too many. What makes it different is that I speak as a planetary physician whose patient, the living Earth, complains of fever; I see the Earth's declining health as our most important concern, our very lives depending upon a healthy Earth. Our concern for it must come first, because the welfare of the burgeoning masses of humanity demands a healthy planet.

At this point my friends and colleagues will wince and wish that I would give up talking of our planet as a form of life.† I understand their concern but I am unrepentant; had I not first thought of the Earth this way we might all have remained 'scientifically correct' but lacked enlightenment about its true nature. Thanks to the concept of Gaia we now see that our planet is entirely different from its dead siblings Mars and Venus. Like one of us, it controls its temperature and composition so as always to be comfortable, and it has done this ever since life began over three billion years ago. To put it bluntly, dead planets are like stone statues, which if put in an oven and heated to 80°C remain unchanged. I would die and so would you if heated that hot, and so would the Earth.

Only when we think of our planetary home as if it were alive can we see, perhaps for the first time, why farming abrades the living tissue of its skin and why pollution is poisonous to it as well as to us. Increasing levels of carbon dioxide and methane gas in the atmosphere have consequences quite different from those that would occur on a dead planet like Mars. The living Earth's response to what we do will depend not merely on the extent of our land use and pollutions but also on its current state of health. When the Earth was young and strong, it resisted adverse change and the failure of its own temperature regulation; now it may be elderly and less resilient.

Sustainable development, supported by the use of renewable energy,† is the fashionable approach to living with the Earth, and it is the platform of green-thinking politicians. Opposing this view, particularly in the United States, are the many who still regard global warming as a fiction and favour business as usual. Their thinking is well expressed in the recent novel by Michael Crichton, *State of Fear*, and by that saintly woman, Mother Theresa, who in 1988 said, 'Why should we care about the Earth when our duty is to the poor and the sick among us. God will take care of the Earth.' In truth, neither faith in God nor trust in business as usual, nor even commitment to sustainable development, acknowledges our true dependence; if we fail to take care of the Earth, it surely will take care of itself by making us no longer welcome. Those with faith should look again at our Earthly home and see it as a holy place, part of God's creation, but something that we have desecrated. Anne Primavesi's book *Gaia's Gift* shows the way to consilience† between faith and Gaia.

When I hear the phrase 'sustainable development' I recall the definition given by Gisbert Glaser, the senior advisor to the International Council for Science, who said in a guest editorial of the International Geosphere Biosphere Program (IGBP) newsletter, 'Sustainable development is a moving target. It represents the continuous effort to balance and integrate the three pillars of social well-being, economic prosperity and environmental protection for the benefit of present and future generations.' Many consider this noble policy morally superior to the *laissez faire* of business as usual. Unfortunately for us, these wholly different approaches, one the expression of international decency, the other of unfeeling market forces, have the same outcome: the probability of disastrous global change. The error they share is the belief that further development is possible and that the Earth will continue, more or less as now, for at least the first half of this century. Two hundred years ago, when change was slow or non-existent, we might have had time to establish sustainable development, or even have continued for a while with business as usual, but now is much too late; the damage has already been done. To expect sustainable development or a trust in business as usual to be viable policies is like expecting a lung cancer victim to be cured by stopping smoking; both measures deny the existence of the Earth's disease, the fever brought on by a plague of people. Despite their difference, they come from religious and humanist beliefs which regard the Earth as there to be exploited for the good of humankind. When there were only one billion of us in 1800, these ignorant policies were acceptable because they caused little harm. Now, they travel two different roads that will soon merge into a rocky path to a Stone Age existence on an ailing planet, one where few of us survive among the wreckage of our once biodiverse Earth.

Why are we so slow, especially in the United States, to see the great peril that faces us and civilization? What stops us from realizing that the fever of global heating is real and deadly and might already have moved outside our and the Earth's control? I think that we reject the evidence that our world is changing because we are still, as that wonderfully wise biologist E. O. Wilson reminded us, tribal carnivores. We are programmed by our inheritance to see other living things as mainly something to eat, and we care more about our national tribe than anything else. We will even give our lives for it and are quite ready to kill other humans in the cruellest of ways for the good of our tribe. We still find alien the concept that we and the rest of life, from bacteria to whales, are parts of the much larger and diverse entity, the living Earth.

Science is supposed to be objective, so why has it failed to warn us sooner of these dangers? Global heating was lightly discussed by several authors in the mid twentieth century, but even that great climatologist Hubert Lamb, in his 1972 book *Climate: Present, Past and Future*, had only one page on the greenhouse effect† in a work covering 600 pages. The subject did not go public until about 1988; before that, most atmospheric scientists were so absorbed by the intriguing science of stratospheric ozone depletion that they had little time for other environmental problems. Among the brave pioneers of the larger issues of global change were the American scientists Stephen Schneider and Jim Hansen. I first met Schneider in the late 1970s during a visit to the National Center for Atmospheric Research, an entrancing place of science perched on a mountainside at Boulder in Colorado, and our paths through science have been interlaced ever since. In his book with Randi Londer, *The Coevolution of Climate and Life*, published in 1984, he warns of the probable consequences of continuing to burn fossil fuels and recommends the need for a strategic control of emissions, not the business as usual of market forces. Jim Hansen of the NASA Goddard Institute of Space Studies was equally strong in his warnings, and on 23 June 1988 he told the United States Senate that the Earth was now warmer than at any time in the history of instrumental measurements. The best and most complete histories of this period are in John Gribbin's book *Hothouse Earth*, published in 1990, Schneider's 1989 book, *Global Warming*, and Fred Pearce's *Turning up the Heat*, published in 1989.

Schneider's and Hansen's words were amplified by politicians as far apart as Al Gore and Margaret Thatcher, and I suspect that credit for their transformation into practical action should go to the diplomat climatologist Sir Crispin Tickell. These considerable efforts led to the formation in 1989, by the World Meteorological Organisation (WMO) and the United Nations Environment Programme (UNEP) under the chairmanship of Professor Bert Bolin, of the Intergovernmental Panel on Climate Change (IPCC). It soon started the long process of data gathering and model building that was the basis for forecasts of future climates. But somehow the sense of urgency about global heating faded in the 1990s, and the pioneering bravery of the whistle blowers received little support from the lumpen middle management of science. They were not wholly to blame, for science itself was handicapped in the last two centuries by its division into many different disciplines, each limited to seeing only a tiny facet of the planet, and there was no coherent vision of the Earth. Scientists did not acknowledge the Earth as a self-regulating entity until the Amsterdam Declaration in 2001, and many of them still act as if our planet were a large public property that we own and share. They cling to their nineteenth- and twentieth-century view of the Earth that was taught at school and university, of a planet made of dead inert rock with abundant life aboard, passengers on its journey through space and time.

Science is a cosy, friendly club of specialists who follow their numerous different stars; it is proud and wonderfully productive but never certain and always hampered by the persistence of incomplete world views. We are fortunate in Britain to have had our science led by those towering figures Lord May and Sir David King, both of whom have tirelessly used their strength to warn us and the government of the huge dangers that loom ahead. The notion of Gaia, with its implication of the Earth as an evolving system that was in some ways alive, did not appear until about 1970. Like all new theories it took decades before it was even partially accepted, because it had to wait for evidence to confirm or deny it. We know now that the Earth really does regulate itself, but because of the time it took to gather the evidence we discovered too late that the regulation was failing and the Earth system was fast approaching the critical state that puts all life on it in danger.

Science tries to be global and more than a loose collection of separated disciplines, but even those who take a systems-science approach would be the first to admit that our understanding of the Earth system is not much better than a nineteenth-century physician's understanding of a patient. But we are sufficiently aware of the physiology of the Earth to realize the severity of its illness. We suspect the existence of a threshold, set by the temperature or the level of carbon dioxide in the air; once this is passed nothing the nations of the world do will alter the outcome and the Earth will move irreversibly to a new hot state. We are now approaching one of these tipping points, and our future is like that of the passengers on a small pleasure boat sailing quietly above the Niagara Falls, not knowing that the engines are about to fail.

The few things we do know about the response of the Earth to our presence are deeply disturbing. Even if we stopped immediately all further seizing of Gaia's land and water for food and fuel production and stopped poisoning the air, it would take the Earth more than a thousand years to recover from the damage we have already done, and it may be too late even for this drastic step to save us. To recover, even to lessen the consequences of our past errors, will take an extraordinary degree of international effort and a carefully planned sequence for replacing fossil carbon with safer energy sources. We as a civilization are all too much like someone addicted to a drug that will kill if continued and kill if suddenly withdrawn. We are in our present mess through our intelligence and inventiveness. It could have started as long as 100,000 years ago, when we first set fire to forests as a lazy way of hunting. We had ceased to be just another animal and begun the demolition of the Earth. We are the species equivalent of that schizoid pair, Mr Hyde and Dr Jekyll; we have the capacity for disastrous destruction but also the potential to found a magnificent civilization. Hyde led us to use technology badly; we misused energy and overpopulated the Earth, but we will not sustain civilization by abandoning technology. We have instead to use it wisely, as Dr Jekyll would do, with the health of the Earth, not the health of people, in mind. This is why it is much too late for sustainable development; what we need is a sustainable retreat.

We are so obsessed with the idea of progress and with the betterment of humanity that we regard retreat as a dirty word, something to be ashamed of. The philosopher and historian of ideas John Gray observed in his book *Straw Dogs* that only rarely do we see beyond the needs of humanity, and he linked this blindness to our Christian and humanist infrastructure. It arose 2,000 years ago and was then benign, and we were no significant threat to Gaia. Now that we are over six billion hungry and greedy individuals, all aspiring to a first-world lifestyle, our urban way of life encroaches upon the domain of the living Earth. We are taking so much that it is no longer able to sustain the familiar and comfortable world we have taken for granted. Now it is changing, according to its own internal rules, to a state where we are no longer welcome.

Humanity, wholly unprepared by its humanist traditions, faces its greatest trial. The acceleration of the climate change now under way will sweep away the comfortable environment to which we are adapted. Change is a normal part of geological history; the most recent was the Earth's move from the long period of glaciation to the present warmish interglacial. What is unusual about the coming crisis is that we are the cause of it, and nothing so severe has happened since the long hot period at the start of the Eocene, fifty-five million years ago, when the change was larger than that between the ice age and the nineteenth century and lasted for 200,000 years.

The great Earth system, Gaia, when in an interglacial period as it is now, is trapped in a vicious cycle of positive feedback,[†] and this is what makes global heating so serious and so urgent. Extra heat from any source, whether from greenhouse gases, from the disappearance of Arctic ice and the changing structure of the ocean, or from the destruction of tropical forests, is amplified and the effects are more than additive. It is almost as if we had lit a fire to keep warm and failed to notice, as we piled on fuel, that the fire was out of control and the furniture had ignited. When that happens there is little time left to put out the fire before it consumes the house itself. Global heating, like a fire, is accelerating and there is almost no time left to act.

The philosopher Mary Midgley, in her splendid books *Science and Poetry* and *The Essential Mary Midgley*, has warned that the dominance of atomistic and reductionist thinking in science during the past two centuries has led to a narrow parochial view of the Earth. We often say in science that eminence is measured by the length of time progress is held up by a scientist's ideas. It took nearly 200 years for Newton's view of the Universe to give way to Einstein's more complete vision. By this measure of eminence, Descartes was a truly great thinker. His separation of mind from body, necessary at the time, and the relegation of all things living to mechanistic interpretation encouraged reductionist thinking. Reduction is the analytical dissection of a thing into its ultimate component parts, followed by regeneration through the reassembly of the parts; it certainly led to great triumphs in physics and biology during the past two centuries, but it is only now falling into its proper place as a part and not the whole of science. At last, but maybe too late, we begin to see that the top-down holistic view, which views a thing from outside and asks it questions while it works, is just as important as taking the thing to pieces and reconstituting it from the bottom up. This is especially true of living things, large systems and computers.

We need most of all to renew that love and empathy for nature that we lost when we began our love affair with city life. Socrates was probably not the first to say that nothing interesting happens outside the city walls, but he would have been familiar with the natural world outside. Even in Shakespeare's time cities were small enough for him to walk to 'a bank whereon the wild thyme blows, where oxlips and the nodding violet grows'. The early environmentalists who knew and truly appreciated nature – Wordsworth, Ruskin, Rousseau, Humboldt, Thoreau and so many others – lived for much of their lives in small compact cities. Now, the city is so huge that few ever experience the countryside, it is so distant. I wonder how many of you know what an oxlip looks like or have seen one.

Blake saw the menace of dark satanic mills, but I doubt if even his worst nightmare vision would have encompassed today's reality, the wholesale industrialization of the countryside he knew. Blake was a Londoner, but from his London, a perfect countryside was no more than a walk away. They no longer make hay in England's green and pleasant land, they farm by mechanized agribusiness; and if we allow it, the remaining countryside will become an industrial site filled with massive wind turbines in a vain attempt to supply the energy demands of urban life. Reform is all too often organized vandalism in the name of ideology. This marred Cromwell's government, and is now the dark side of European green politics.

Of course there are sceptics, and among them are the Danish statistician Björn Lomborg and the American scientist Richard Lindzen, both of whom doubt that global change is anywhere near so large a problem that we need do anything about it now. These contrary views have not swayed the consensus of the many scientists from around the world who form the IPCC.

Recently I listened to a passionate and moving speech broadcast by the American scientist Patrick Michaels. He indignantly rejected the claim by Sir David King, the United Kingdom's chief scientific adviser, that global warming was more serious than the war now being waged against terrorism. To him, and many others, the events of 11 September 2001, Madrid 2004, and London 2005 far transcend in importance remote forecasts of bad weather in the coming century. Unlike most Americans, I have spent most of my lifetime under the threat of terrorism, which came mostly but not exclusively from Celtic nationalism. I share Michaels' indignation and regard terrorism as but one level less evil than genocide. Terrorism and genocide both result from our tribal natures. Tribal behaviour is surely written in the language of our genetic code, or why else would we as a mob or a crowd do the evil things that only psychopaths would do alone. Genocide and terrorism are not the singular evils of our enemies; all of us are capable given the right signal, and civilization has only slightly sanitized these awful trends and called them war. Tribalism is not wholly bad and can be mobilized to make us otherwise selfish humans perform truly bravely and even give our lives, usually because we sense a danger to our tribe but sometimes for the good of humankind. We do remarkably good things unselfishly. In wartime we accept severe rationing of food and consumer goods; we willingly work for longer hours and face great danger, and some even eagerly face death.

I am old enough to notice a marked similarity between attitudes over sixty years ago towards the threat of war and those now towards the threat of global heating. Most of us think that something unpleasant may soon happen, but we are as confused as we were in 1938 over what form it will take and what to do about it. Our response so far is just like that before the Second World War, an attempt to appease. The Kyoto agreement was uncannily like that of Munich, with politicians out to show that they do respond but in reality playing for time. Because we are tribal animals, the tribe does not act in unison until a real and present danger is perceived. This has not yet happened; consequently, as individuals, we go our separate ways while the ineluctable forces of Gaia marshal against us. Battle will soon be joined, and what we now face is far more deadly than any blitzkrieg. By changing the environment we have unknowingly declared war on Gaia. We have infringed the environment of the other species, just as if, in the affairs of nation states, we had occupied the land of other nations.

The prospects are grim, and even if we act successfully in amelioration, there will still be hard times, as in any war, that will stretch us to the limit. We are tough and it would take more than the predicted climate catastrophe to eliminate all breeding pairs of humans; what is at risk is civilization. As individual animals we are not so special, and in some ways the human species is like a planetary disease, but through civilization we redeem ourselves and have become a precious asset for the Earth. There is a small chance that the sceptics are right, or we might be saved by an unexpected event such as a series of volcanic eruptions severe enough to block out sunlight and so cool the Earth. But only losers would bet their lives on such poor odds. Whatever doubts there are about future climates, there are no doubts that both greenhouse gases and temperatures are rising.

I find it sad and ironic that the United Kingdom, which leads the world in the quality of its Earth and climate scientists, has rejected their warnings and advice. We have so far preferred to listen to the well-intended but unwise advice of those who think there is an alternative to science. I am a green and would be classed among them, but I am most of all a scientist; because of this I entreat my friends among greens to reconsider their naive belief in sustainable development and renewable energy, and that this and saving energy are all that need be done. Most of all, they must drop their wrongheaded objection to nuclear energy. Even if they were right about its dangers, and they are not, its use as a secure, safe and reliable source of energy would pose a threat insignificant compared with the real threat of intolerable and lethal heatwaves and sea levels rising to threaten every coastal city of the world. Renewable energy sounds good, but so far it is inefficient and expensive. It has a future, but we have no time now to experiment with visionary energy sources: civilization is in imminent danger and has to use nuclear energy now, or suffer the pain soon to be inflicted by our outraged planet. We must follow the good green advice to save energy, and we must all do this whenever we can, but I suspect that, like losing weight, it is easier said than done. Significant energy saving comes from improved designs, and these take decades to reach the majority of users.

I am not recommending nuclear fission energy as the long-term panacea for our ailing planet or as the answer to all our problems. I see it as the only effective medicine we have now. When one of us develops late-onset diabetes as a consequence of overeating and insufficient exercise, we know that medicine alone is not enough; we have to change our whole style of living. Nuclear energy is merely the medicine that sustains a steady secure source of electricity to keep the lights of civilization burning until clean and everlasting fusion, the energy that empowers the sun, and renewable energy are available. We will have to do much more than just rely on nuclear energy if we are to avoid a new Dark Age later in this century.

We must conquer our fears and accept nuclear energy as the one safe and proven energy source that has minimal global consequences. It is now as reliable as any human engineering can be and has the best safety record of all large-scale energy sources. France has shown that it can become a major national source of energy, yet governments are still fearful of grasping this one lifeline we can use immediately. We need a portfolio of energy sources, with nuclear playing a major part, at least until fusion power becomes a practical option. If food can be synthesized by the chemical and biochemical industries from carbon dioxide, water and nitrogen, then let's make it and give the Earth a rest. We must stop fretting over the minute statistical risks of cancer from chemicals or radiation. Almost a third of us will die of cancer anyway, mainly because we breathe air laden with that all pervasive carcinogen, oxygen. If we fail to concentrate our minds on the real danger, which is global heating, we may die even sooner, as did more than 30,000 unfortunates from overheating in Europe in the summer of 2003. We have to take global change seriously and immediately and then do our best to lessen the footprint of humans on the Earth. Our goal should be the cessation of fossil-fuel consumption as quickly as possible, and there must be no more natural-habitat destruction anywhere. When I use the term 'natural' I am not thinking only of primeval forests: I include also the forests that have grown back when farmland was abandoned, as happened in New England and other parts of the USA. These re-established forests probably perform their Gaian services as well as did the original forests, but the vast open stretches of monoculture farmland are no substitute for natural ecosystems. We are already farming more than the Earth can afford, and if we attempt to farm the whole Earth to feed people, even with organic farming, it would make us like sailors who burnt the timbers and rigging of their ship to keep warm. The natural ecosystems[†] of the Earth are not just there for us to take as farmland; they are there to sustain the climate and the chemistry of the planet.

To undo the harm we have already done requires a programme whose scale dwarfs the space and military programmes, in cost and size. We live at a time when emotions and feelings count more than truth, and there is a vast ignorance of science. We have allowed fiction writers and green lobbies to exploit the fear of nuclear energy and of almost any new science, in the same way that the churches exploited the fear of Hellfire not so long ago. We are like passengers on a large aircraft crossing the Atlantic Ocean who suddenly realize just how much carbon dioxide their plane is adding to the already overburdened air. It would hardly help if they asked the captain to turn off the engines and let the plane travel like a glider by wind power alone. We cannot turn off our energy-intensive, fossil-fuel-powered civilization without crashing; we need the soft landing of a powered descent.

The time of irreversible adverse change may be so close that it would be unwise to rely on international agreement to save civilization from the consequences of global heating. The G8 meeting in Scotland in 2005 had climate change as an agenda item but it was marginalized when London experienced a serious terrorist incident. We cannot afford to wait for Godot. Without losing sight of the global scale of the danger, individual nations may need to think of ways to save themselves as well as the world. We in the UK are as we were in 1939 and may soon be, to a considerable extent, alone; our future food and energy supplies can no longer be taken as secure from a world that is devastated by climate change. We have to make decisions based on our national interest. This is neither chauvinist nor selfish: it could be the fastest way to ensure that more and more nations, driven by their own self-interest, act locally over global change. The large emergent nations, India and China, will find it difficult to rein in their use of fossil fuel, as will the USA. We should not wait for international agreement or instruction.

In our small country we have to act now as if we were about to be attacked by a powerful enemy. We have first to make sure our defences against climate change are in place before the attack begins. The most vulnerable places are the cities close to sea level now, and among them are London and Liverpool. First we need to ensure that they are adequately defended for the early stages of the climate war and then be prepared to retreat from them in an orderly way as the floods advance. Once the Earth begins to move rapidly to its new hotter state, climate change will surely disrupt the political and trading world. Imports of food, fuels and raw materials will increasingly become inadequate as the suppliers in other regions are overwhelmed by droughts and floods. We need to plan for the synthesis of food from nothing more than air, water and a few minerals, and this will require a secure and abundant source of energy. The highly productive farmlands of eastern England will be among the first areas to be inundated.

The only sources of energy we can rely on will be coal, the little that remains of North Sea oil and gas, nuclear energy and a small amount of renewable energy. The extravagant and intrusive building of on-shore wind farms should cease immediately and the funds released be used for practical renewable energy schemes such as the Severn Estuary tidal barrage; this might provide a steady 5 to 10 per cent of the energy needs of our nation when we stop the present wasteful misuse. We need, most of all, that change of heart and mind that comes to tribal nations when they sense real danger. Only then will we accept the hardships of fuel rationing and firm constraints that an effective defence demands. Our cause will be the defence of our civilization to ward off the chaos that might otherwise overtake us.

Astronauts who have had the chance to look back at the Earth from space have seen what a stunningly beautiful planet it is, and they often talk of the Earth as home. I ask that we put our fears and our obsession with personal and tribal rights aside, and be brave enough to see that the real threat comes from the harm we do to the living Earth, of which we are a part and which is indeed our home.

2

What is Gaia?

Hardly anyone, and that included me for the first ten years after the concept was born, seems to know what Gaia is. Most scientists, when they think and talk about the living part of the Earth, call it the biosphere,[†] although strictly speaking the biosphere is no more than the geographical region where life exists, the thin spherical bubble at the Earth's surface. They have unconsciously expanded the definition of the biosphere into something larger than a geographical region but seem vague about where it starts and ends geographically and what it does.

Going outwards from the centre, the Earth is almost entirely made of hot or molten rock and metal. Gaia is a thin spherical shell of matter that surrounds the incandescent interior; it begins where the crustal rocks meet the magma of the Earth's hot interior, about 100 miles below the surface, and proceeds another 100 miles outwards through the ocean and air to the even hotter thermosphere at the edge of space. It includes the biosphere and is a dynamic physiological system that has kept our planet fit for life for over three billion years. I call Gaia a physiological system because it appears to have the unconscious goal of regulating the climate and the chemistry at a comfortable state for life. Its goals are not set points but adjustable for whatever is the current environment and adaptable to whatever forms of life it carries.

We have to think of Gaia as the whole system of animate and inanimate parts. The burgeoning growth of living things enabled by sunlight empowers Gaia, but this wild chaotic power is bridled by constraints which shape the goal-seeking entity that regulates itself on life's behalf.

I see the recognition of these constraints to growth as essential to the intuitive understanding of Gaia. Important to this understanding is that constraints affect not only the organisms or the biosphere but also the physical and chemical environment. It is obvious that it can be too hot or too cold for mainstream life, but not so obvious is the fact that the ocean becomes a desert when its surface temperature rises above about 12°C; when this happens, a stable surface layer of warm water forms that stays unmixed with the cooler, nutrient-rich waters below. This purely physical property of ocean water denies nutrients to the life in the warm layer, and soon the upper sunlit ocean water becomes a desert. This may be one of the reasons why Gaia's goal appears to be to keep the Earth cool.

You will notice I am continuing to use the metaphor of 'the living Earth' for Gaia; but do not assume that I am thinking of the Earth as alive in a sentient way, or even alive like an animal or a bacterium. I think it is time we enlarged the somewhat dogmatic and limited definition of life as something that reproduces and corrects the errors of reproduction by natural selection among the progeny.

I have found it useful to imagine the Earth as like an animal, perhaps because my first experience of serious science as a graduate was in physiology. It has never been more than metaphor – an *aide pensée*, no more serious than the thoughts of a sailor who refers to his ship as 'she'. Until recently no specific animal came into my mind, but always something large, like an elephant or a whale. Recently, on becoming aware of global heating, I have thought of the Earth more as a camel. Camels, unlike most animals, regulate their body temperatures at two different but stable states. During daytime in the desert, when it is unbearably hot, camels regulate close to 40°C, a close enough match to the air temperature to avoid having to cool by sweating precious water. At night the desert is cold, and even cold enough for frost; the camel would seriously lose heat if it tried to stay at 40°C, so it moves its regulation to a more suitable 34°C, which is warm enough. Gaia, like the camel, has several stable states so that it can accommodate to the changing internal and external environment. Most of the time things stay steady; as they were over the few thousand years before about 1900. When the forcing is too strong, either to the hot or the cold, Gaia, as a camel would, moves to a new stable state that is easier to maintain. She is about to move now.

Metaphor is important because to deal with, understand, and even ameliorate the fix we are now in over global change requires us to know the true nature of the Earth and imagine it as the largest living thing in the solar system, not something inanimate like that disreputable contraption 'spaceship Earth'. Until this change of heart and mind happens we will not instinctively sense that we live on a live planet that can respond to the changes we make, either by cancelling the changes or by cancelling us. Unless we see the Earth as a planet that behaves as if it were alive, at least to the extent of regulating its climate and chemistry, we will lack the will to change our way of life and to understand that we have made it our greatest enemy. It is true that many scientists, especially climatologists, now see that our planet has the capacity to regulate its climate and chemistry, but this is still a long way from being the conventional wisdom. It is not easy to grasp the concept of Gaia, a planet able to keep itself fit for life for a third of the time the universe has existed, and until the IPCC sounded the alarm there was little inclination. I will try to provide an explanation that would satisfy a practical person like a physician. A complete explanation that would satisfy a scientist may be inaccessible, but the lack of it is no excuse for inaction.

I find explaining Gaia is like teaching someone how to swim or to ride a bicycle: there is much that cannot be put into words. To make it easier I will start at the shallow end with a simple question that illustrates the mind-wrenching difference between two equally important ways of thinking about the world. The first is systems science, which is about anything alive, whether an organism or an engineering mechanism while it is working; the second is reductionist science, the cause-and-effect thinking that has dominated the last two centuries of science. The question is: what has peeing to do with the selfish gene?

When I was a young man I was amazed by the number of euphemisms that existed for the simple but essential practice of passing urine. Doctors and nurses would ask you to 'produce a specimen' or 'pass some water' and often hand out a small container to make their request clear. In everyday speech we 'pumped the ship', 'sprung a leak' or 'shed the load' and we did it in 'the little boys' room' or the 'bathroom'. Sometimes we just 'spent a penny'.

Perhaps it was all a hangover from the nineteenth-century confusion over sex. It was not only impossible in polite speech to mention the genitals; the taboo applied also to their alternative uses. But as the outstanding American biologist George Williams observed in 1996, what an odd evolutionary economy to use the same organ for pleasure, reproduction and waste disposal. It was not until quite recently that I began to wonder if there might not be something deeper lurking behind this minor mystery. Why do we pee? Not so silly a question as it might seem. The need to rid oneself of waste products like excess salt, urea, creatinine and numerous other scraps of metabolism is obvious but only part of the answer. Perhaps we pee for altruistic reasons. If we and other animals did not pass urine some of the vegetable life of the Earth might be starved of nitrogen.

Is it possible that in the evolution of Gaia, the great Earth system, animals have evolved to excrete nitrogen as urea or uric acid instead of gaseous nitrogen? For us the excretion of urea represents a significant waste of energy and of water. Why should we evolve something to our disadvantage unless it was for altruistic reasons? Urea is the waste product of the metabolism of the meat, the fish, the cheese and the beans we eat; all are rich in protein, the stuff of life. We digest what we eat and break it down to its component chemicals; we do not take beef muscle protein and use it in our own muscles. We build or replace our muscles and other tissue by assembling the component parts, the amino acids of the proteins, into fresh protein according to the plan in our DNA. To use the protein from beef directly to make our muscles would be like taking the parts of a tractor to repair a washing machine. The waste left over from this busy construction and deconstruction ultimately becomes urea, and we seem to have no option but to get rid of it as a dilute solution in water, urine.

Urea is a simple chemical, a combination of ammonia and carbon dioxide, or as an organic chemist would say, the di-amide of carbonic acid, NH_2CONH_2. Why did we and other mammals evolve to excrete our nitrogen in this form? Why not break down the urea into carbon dioxide, water and nitrogen gas? Much easier to excrete nitrogen by breathing it out, and it would save the water needed for excreting urea; oxidizing the urea would even add a little water, to say nothing of providing more energy.

Let us look at the figures. 100 grams of urea is metabolically worth 90 kilocalories or, if you prefer, 379 kilojoules. But if instead of being consumed it is passed in urine, more than four litres of water are needed to excrete the 100 grams of urea at a non-toxic dilution. Normally we excrete about 40 grams of urea daily in about 1.5 litres of water. Not much of a problem, you might think, but just consider animals living in a desert region short of food and water. If a mutant appeared that was able to metabolize urea to nitrogen, carbon dioxide and water, it would be at a considerable advantage and probably be able to leave more progeny than its urea-excreting competitors. According to a simplistic interpretation of Darwinian theory, selection would favour this mutant trait and it would spread rapidly and become the norm.

At this point a sceptical biochemist will say, 'Don't you realize that the products of ammonia or urea oxidation are all poisonous, and that is why we excrete nitrogen as urea?' My reply would be, 'Tell that to the bacteria that change nitrogen compounds into nitrogen gas and which are abundant in the soil and ocean.' More than this, a symbiosis with denitrifying organisms might be as good as or better than trying to metabolize urea ourselves.

So you see, urea is waste for us and wasting it loses valuable water and energy. But if we and other animals did not pee and breathed out nitrogen instead, there might be fewer plants and later we would be hungry. How on Earth did we evolve to be so altruistic and have such enlightened self interest? Perhaps there is wisdom in the workings of Gaia and the way she interprets the selfish gene.

When I started working on Gaia forty years ago, science was not as now a highly organized and often corporate enterprise. There was almost no forward planning or status reports, and there were almost never meetings to plan what to do next. There was no health and safety bureaucracy – we were expected to be, as qualified scientists, responsible for our own and our colleagues' safety. Most differently, science was done hands-on in the laboratory, not simulated on a computer screen in an office or a cubicle. In this idyllic environment it was possible to do an experiment to confirm or deny an idea. Sometimes the answer was a simple right or wrong, but on other occasions something equivocal. These 'don't knows' were what led by serendipity to the revelation of something wholly unexpected, a real discovery.

So it might be with the idea of urea excretion. Thinking about nitrogen this way led me to wonder about the vexing problem of oxygen in the Carboniferous period some 300 million years ago. An important part of the evidence for Gaia comes from the abundance of atmospheric gases, such as oxygen and carbon dioxide; these are regulated at a level comfortable for whatever happens to be the current form of life. There are good experimental as well as theoretical grounds for thinking that the present percentage of oxygen in the atmosphere is about right. More than 21 per cent carries an increasing fire risk; at 25 per cent the probability of a blaze from a spark increases about tenfold. Andrew Watson and Tim Lenton have modelled the regulation of oxygen and have found the fire risk of dry vegetation to play an important part in the mechanism of oxygen regulation. Below 13 per cent there are no fires, and above 25 per cent they are so fierce that it seems impossible that forests could reach maturity. Imagine our surprise when the eminent geochemist Robert Berner proposed that in the Carboniferous period, about 300 million years ago, oxygen was 35 per cent of the atmosphere. His conclusion came from a model based on a thorough analysis of the composition of carboniferous rocks. He argued that at that time so much carbon was being buried, much of which we now see as the coal measures, that there had to be much more oxygen in the air to balance this greater rate of carbon burial.

My first reaction was that Berner must be wrong; I knew from the careful experiments made by my colleague Andrew Watson in the 1970s that fires in 35 per cent oxygen are almost as fierce as in pure oxygen. I was not impressed by laboratory experiments that suggested that twigs from trees did not readily inflame in 35 per cent oxygen; there is a world of difference between a laboratory simulation and a real forest fire, where its intense radiation dries out the wood in the path of the fire and where the winds drawn by the fire bring in fresh oxygen-rich air. Nor was I impressed by arguments that the huge dragonflies that existed at that time could not have flown without 35 per cent oxygen in the air. It is now realized that insects are unusually vulnerable to oxygen poisoning and that the Cretaceous dragonflies would have had no difficulty flying at our present oxygen levels. The argument went on until a friend, Andrew Thomas, an acoustic scientist and also a diver, suggested that maybe we were both right. Berner was right to claim that there was more oxygen and I was right to say it could not have been present at much over 25 per cent. All that was required was more nitrogen in the air. It is not the amount of oxygen that determines flammability, but its proportion in the mixture with nitrogen.

About 40 per cent of the nitrogen on Earth is now buried in the crust; perhaps in the Cretaceous that nitrogen had not yet been buried and existed in the air and so kept the proportion of oxygen safer for trees. We might also speculate that the microbial life of the Precambrian that preceded the appearance of trees and animals did not conserve nitrogen, so that it would have been present mainly as gas in the air.

These thoughts about nitrogen are wholly speculative, but I include them to illustrate the way that Gaia theory[†] has developed from ideas that were at first vague or from fruitful errors that were the seeds from which a truer account has emerged.

So let us go deeper now and try to sense Gaia by looking at the Earth from outside as a whole planet. Imagine a spacecraft manned by intelligent aliens who are looking at the solar system from space. They would have aboard their ship instruments powerful enough to show the travellers the chemical composition of every planet's atmosphere. From this analysis and nothing more, their automated instruments would tell them that the only planet with abundant life was the Earth; more than that, they would say that the life form was carbon-based and was sufficiently advanced to have an industrial civilization. There is nothing science fictional about the instrument itself; a small telescope with an infra red spectrometer and a computer to control them and analyse their observations would do. They would see methane and oxygen coexisting in the upper air of the Earth, and the ship's scientist would know that these gases were reacting in the bright sunlight and that therefore something on the ground must be making large quantities of them both. The odds against this happening by chance inorganic chemistry are near infinity. They would conclude that our planet is a rich habitat for life, and the presence of CFCs would suggest a civilization unwise enough to have allowed their escape.

In the 1960s I was a contractor designing instruments for NASA's planetary exploration team, and thoughts like these led me to propose planetary atmospheric analysis for the detection of life on Mars. I argued that if there was life on Mars it would have to use the atmosphere as a source of raw materials and as somewhere to deposit its wastes; this would change the atmospheric composition and make it recognizably different from that of a dead planet. I saw the Earth, rich with life, as the contrasting planet, and I used the eminent scientist G. E. Hutchinson's authoritative review of biogeochemistry as my source of information on the sources and sinks for the gases of the air. He reported methane and nitrous oxide as biological products, and nitrogen, oxygen and carbon dioxide as massively changed in abundance by organisms. At the time, none of us knew much about the composition of Mars's atmosphere, but in 1965 Earth-based infra-red astronomy revealed the Mars atmosphere to be composed almost entirely of carbon dioxide and close to chemical equilibrium; according to my proposal it was therefore probably lifeless – not a popular conclusion to give my sponsors. Turning aside from life detection, I wondered what could be keeping our chemically unstable atmosphere in dynamic steady state and the Earth always apparently habitable. Moreover, the continuity of life requires a tolerable climate despite a 37 per cent increase of solar luminosity since the Earth formed. Together, these thoughts led me to the hypothesis that living organisms regulate the climate and the chemistry of the atmosphere in their own interest, and in 1969 the novelist William Golding proposed Gaia as its name. A few years later, I started collaborating with the eminent American biologist Lynn Margulis, and in our first joint paper we stated: the Gaia hypothesis views the biosphere as an active, adaptive control system able to maintain the Earth in homeostasis.

From its beginning in the 1960s, the idea of the global self-regulation of climate and chemistry was unpopular with both Earth scientists and life scientists. At best, they found it unnecessary as an explanation of the facts of life and the Earth; at worst, they condemned it outright in scathing terms. The only scientists who welcomed the idea were a few meteorologists and climatologists. Some biologists soon challenged the hypothesis, arguing that a self-regulating biosphere could never have evolved, since the organism was the unit of selection, not the biosphere. I was fortunate to have that fine and clear author Richard Dawkins as the advocate for the Darwinian opposition to Gaia; it was painful but in time I found myself agreeing with him that Darwinian evolution, as it was then understood, was incompatible with the Gaia hypothesis.† I did not doubt Darwin, so what was wrong with the Gaia hypothesis? I knew that the constancy of climate and of the chemical composition of the air were good evidence for a self-regulating planet. Moreover, the concept of Gaia is fruitful, and it led me to discover the natural molecular carriers of the elements sulphur and iodine: dimethyl sulphide (DMS) and methyl iodide. Several years later in 1986, while collaborating with colleagues in Seattle, we made the awesome discovery that DMS from ocean algae† was connected with the formation of clouds and with climate. We were moved to catch a glimpse of one of Gaia's climate-regulation mechanisms, and we were indebted to the climate-science community who took us seriously enough to award to the four of us, Robert Charlson, M. O. Andreae, Steven Warren and me, their Norbert Gerbier Prize in 1988.

To return to the arguments with the Darwinists, it occurred to me in 1981 that Gaia was the whole system – organisms and material environment coupled together – and it was this huge Earth system that evolved self-regulation, not life or the biosphere alone. To test this idea I composed a computer model of dark- and light-coloured plants competing for growth on a planet in progressively increasing sunlight. It was no more than a simulation of the world, but the running program showed the imaginary world regulating its temperature close to the optimum for daisy growth and over a wide range of heat outputs from its star. This model, which I called Daisyworld, was unusual for an evolutionary model made from coupled differential equations; it was stable, insensitive to initial conditions and resistant to perturbation.

Daisyworld models a planet like the Earth, orbiting a star like our sun. On Daisyworld there are only the two plant species, and they both compete for living space as any plants would do. When the sun is younger and cooler, so is the model planet, and at that time the dark daisies flourish. Only at the hottest places near the equator are light daisies found. This is because dark daisies absorb sunlight and keep themselves, their region and the whole planet warm. As the star heats up, the dark daisies living in the tropics are displaced by light daisies, because the light ones reflect sunlight and so are cooler; they also cool their region and the whole planet. As the star continues to warm, the light daisies displace the dark, and through their competition for space the planet always stays near to the ideal temperature for life. Eventually, the star grows so hot that even light daisies can no longer survive and the planet becomes a lifeless ball of rock.

The model is no more than a caricature, but think of it like that splendid map of the London Tube system – not good as a guide to the streets of London, but ideal for finding your way around the tube system of that bustling city. Daisyworld was invented to show that Darwin's theory of evolution from natural selection is not contrary to Gaia theory, but part of it.

The main reaction of biologists and geologists to Daisyworld was, as good scientists, to try to falsify it, and this they did repeatedly, with increasing irritation, but none succeeded. To answer some of these critics I made models much richer in species than Daisyworld. They included many different types of plant, rabbits to graze them and foxes as predators. They were just as stable and self-regulating as Daisyworld. My friend Stephan Harding has made models of whole ecosystems complete with food webs and used them to enlighten our understanding of biodiversity. The persistence of the critics made me realize that Gaia would not be taken as serious science until eminent scientists approved of it in public. In 1995 I started dialogues with John Maynard Smith and William Hamilton, both of whom were prepared to discuss Gaia as a scientific topic but neither of whom could see how planetary self-regulation could evolve through natural selection. Even so, Maynard Smith gave unstinting support to my friend and colleague Tim Lenton, when the latter wrote a seminal article in *Nature* called 'Gaia and Natural Selection'. In it he described the several ways that the Earth keeps to its goal of sustaining habitability for whatever life forms happen to be its inhabitants. Hamilton wondered in a joint paper with Lenton, with the provocative title 'Spora and Gaia', if the need for organisms to disperse was the link that connected ocean algae with climate. In 1999 Hamilton said in a television programme, 'Just as the observations of Copernicus needed a Newton to explain them, we need another Newton to explain how Darwinian evolution leads to a habitable planet.'

Then, at least in Europe, the ice began to melt, and at a meeting in Amsterdam in 2001 – at which four principal global-change organizations were represented – more than a thousand delegates signed a declaration that had as its first main statement: 'The Earth System behaves as a single, self-regulating system comprised of physical, chemical, biological and human components.'

These words marked an abrupt transition from a previously solid conventional wisdom in which biologists held that organisms adapt to, but do not change, their environments and in which Earth scientists held that geological forces alone could explain the evolution of the atmosphere, crust and oceans. We should recall at this point the trials of that eminent biologist Eugene Odum, who in the 1960s saw an ecosystem as an entity like Gaia. So far as I am aware, none of the biologists who stridently rejected Odum's concept have admitted that they were wrong.

The Amsterdam Declaration was an important step towards the adoption of Gaia theory as a working model for the Earth; however, territorial divisions and lingering doubts kept the declaring scientists from stating the *goal* of the self-regulating Earth, which is, according to my theory, to sustain habitability. This omission allows scientists to pay lip service to Earth System Science (ESS)[†], or Gaia, but continue to model and research in isolation as before. This natural and human tendency of scientists to resist change would not ordinarily have mattered: eventually the strings of habit would have broken and geochemists would have started to think of the biota as an evolving and responding part of the Earth, not as if life were merely a passive reservoir like the sediments or the oceans. Eventually also biologists would have thought of the environment as something that organisms actively changed and not as something fixed to which they adapted. But unfortunately, while scientists are slowly changing their minds, we of the industrial world have been busy changing the surface and atmosphere. Now humanity and the Earth face a deadly peril, with little time left to escape. If the middle management of science had been somewhat less reactionary about Gaia, we might have had twenty more years in which to resolve the much more difficult human and political decisions about our future.

HOW DOES GAIA WORK?

The key to understanding Gaia is to remember that it operates within a set of bounds or constraints. All life is urged by its selfish genes to reproduce, and if the only constraints are competition and predation, the result is a chaotic fluctuation of populations. Attempts to model natural ecosystems that do not include environmental constraints, from the famous rabbits and foxes model of the biophysicist Alfred Lotka and his colleague Vito Volterra, to the latest attempts using complexity theory, all fail to produce the robust stability of a natural ecosystem. Lotka warned as long ago as 1925 that the equations of these too-simple models lacked a constraining physical environment and would be difficult to solve.

In spite of this warning, the abstract mathematics of population biology has fascinated academic biologists for at least seventy years, but it hardly represents the real world, or satisfies their down-to-earth colleagues, the muddy boots ecologists. Examine any long-term natural ecosystem in one of the few remaining untouched places of the Earth, and you will find it is dynamically stable, just like your own body.

Many twentieth-century biologists approached their science with a faith in the infallibility of a genetic description of life. Their faith was so strong that they could not envisage the evolution of an ecosystem happening independently of the genes of its constituent organisms. In fact, the epigenetic evolution of ecosystems and Gaia can take place simply by the selection of existing species. When an ecosystem experiences continued disturbance, such as excessive heat or drought, those species that are tolerant are selected from the ensemble of existing genotypes and they may grow until they dominate; the fine tuning of genetic evolution completes the process of adaptation. The evolution of ecosystems and of Gaia involves more than the selfish gene.

The unstable mathematics of unconstrained competition and predation among living organisms is not unlike the behaviour of the unruly, often drunken, mobs that gather in the city centres at night. The constraint of a strong community confident in its power and backed up by an effective police force once gave quiet and stability, but it has gone and often chaos rules. Gaia itself is firmly constrained by feedback from the non-living environment. Darwinists are right to say that selection favours the organisms that leaves alive the most progeny, but vigorous growth takes place within a constrained space where feedback from the environment allows the emergence of natural self-regulation.

The consequences of unconstrained exponential growth have often been calculated and used as examples of the vigour of life. If a single bacteria divided and repeated that division every twenty minutes, provided that there were no constraints to growth and the food supply was unlimited, in just over two days the total progeny would weigh as much as the Earth. Predation and limits to the supply of nutrients are the local constraints, and pre-Gaia these were all that biologists considered. Now we know that such global properties as atmospheric and oceanic composition and climate set the constraints that bring stability.

So how do these environmental constraints work? They depend upon the tolerances of the organisms themselves. All life forms have a lower, an upper and an optimum temperature for growth, and the same is true for acidity, salinity and the abundance of oxygen in air and water. Consequently, organisms have to live within the bounds of these properties of their environment.

Apart from a few highly adapted organisms, the extremophiles, which live in hot springs near to the boiling point or in the saturated brine of salt lakes or even in the strong acid of our stomachs, almost all life forms are quite fussy about their living conditions. The individual cells that constitute life demand exactly the right mix of salts and nutrients in their internal environment and will tolerate only small changes in the composition of the world around them. When these cells aggregate in their billions to form large animals and plants they can regulate their internal milieu independently of environmental change; we are not harmed by swimming in salt water or by taking a sauna. But bacteria, algae and other single-cell organisms have no choice but to live at whatever temperature and other conditions they find themselves in, and consequently they have adapted to a considerable range of temperature, salinity and acidity. But even for them the temperature range is limited to between −1.6°C, when sea water freezes, to 50°C. We humans and most mammals and birds chose to regulate ourselves close to 37°C and are called homeotherms. The less fussy reptiles and invertebrates are called that curious word poikilotherms or, as we would say, cold blooded. Our own bodies can withstand an internal temperature of 34 or 41°C for short periods, but we are definitely unwell if below 36 or above 39°C. Whether we live as Inuits in the Arctic or as Bushmen in the heat of the Kalahari Desert, those are our internal limits.

Mainstream life flourishes best between 25 and 35°C, but this is only the physiological part of regulation; life is also influenced by the physical properties of the material parts of the Earth. Above 4°C water expands as it warms, and if the ocean surface is warmed from above by sunlight, the top layer absorbs most of the sun's heat and expands to become lighter than the still colder waters beneath. This warmer surface layer has a depth of between 30 and 100 metres. It forms when the sunlight is strong enough to raise the surface temperature above about 10°C.

The warm surface layer is stable, and except in fierce storms, like hurricanes, it stays intact and the cooler waters below do not mix with it. The formation of the surface layer exerts a powerful constraint on ocean life; primary producers that seed the newly formed warm layer in early spring soon go through a succession that uses up nearly all the nutrients of the layer. The dead bodies of this spring bloom sink to the ocean floor, and soon the surface layer is empty of all but a limited and starving population of algae. This is why warm and tropical waters are so clear and blue; they are the deserts of the ocean, and just now they occupy 80 per cent of the world's water surface. In the Arctic and Antarctic, the surface waters remain below 10°C and so are well mixed from the bottom to the surface and nutrients are available everywhere.

In the early part of the twentieth century intercontinental travellers went by sea. Those on a ship travelling to Europe from New York would first see the clear blue warm waters of the Gulf Stream, and then quite suddenly, as they sailed north and east past Cape Cod and entered the Labrador cold current, the water would turn dark and soupy. Ocean life may like to be warm, but the properties of water prevent them from enjoying warmth much above 10°C, unless they are prepared to stay at small numbers and near starvation. This is an important global constraint to growth and is why Gaia does better when cool.

There are oases in the vast deserts of the present world oceans, and these are found at the edges of continents where cold nutrient-rich water wells up from the depths. The seas beyond the estuaries of large rivers like the Mississippi, the Rhine, the Indus and the Yangtze are artificial oases, rich in nutrients, the run-off from intensive agriculture on the land. But these oases, natural and artificial, play only a small part.

A similar and equally important constraint to growth operates on the land surface. Living organisms flourish as it grows warmer up to nearly 40°C, but in the natural world the water they need for life becomes difficult to access once the temperature is much above 20°C. In wintertime when it rains and temperatures are below 10°C, the water stays around for quite a while and the soil stays moist and suitable for growth. In summertime, with average temperatures near 20°C, newly deposited rain soon evaporates and leaves the surface dry; soil loses moisture unless the rain is repeated frequently. Somewhere above 25°C evaporation is so rapid that without continuous rain the soil dries out and the land becomes a desert. Just as in the surface layer of the ocean, organisms may like it warm but the properties of water set a limit to growth.

Richard Betts of the Hadley Centre has shown how the great tropical rainforests have to some extent overcome this limitation by adapting to their warm environment so as to be able to recycle water. The ecosystem does it by sustaining the clouds and rain above the forest canopy, but this ability has its limits. He and Peter Cox suggest a 4°C rise in temperature would be enough to disable the Amazon forest and turn it into scrub or desert, and it would happen partly from the local consequences of a faster evaporation of rain but also from global changes in wind patterns in a 4°C warmer world.

Pure water freezes at 0°C, while in the oceans the salt in the water lowers the freezing point to −1.6°C. Life can adapt to temperatures below freezing – fish swim in water still unfrozen but below 0°C – but active life is impossible in the frozen state. When Sandy and I visited the British Antarctic Survey's labs at Cambridge we were enthralled to see a fish, in a tank held at −1.6°C, swim in a live and responsive way to our host, Lloyd Peck, in anticipation of food. For the fish this was obviously an acceptable temperature. When water is taken from an organism to form ice or as water vapour in drying, the dissolved salts in the organism are concentrated. If the concentration of salt rises above 8 per cent death is immediate. Organisms have adapted to some extent to this problem; sea water, for example, is 6 per cent salt and close to this lethal limit; selection has favoured those organisms that can make substances that neutralize the harmful consequences of increased salt. In the ocean they make large quantities of dimethyl sulphonio propionate (DMSP) for this purpose; on the land insects in the Arctic have evolved antifreeze compounds that prevent salt from accumulating to lethal levels when they freeze.

These physical constraints set by the properties of water feed back on growth and set the shape of the relationship between growth and temperature and the distribution of life on the Earth. From a purely human viewpoint the present interglacial, at least before we started to meddle with it, is a better state than a glaciation. This may be because the more influential humans live in northern hemispheric regions that were either covered in glaciers or tundra during the ice age.

From Gaia's viewpoint the glaciation was a desirable state, with much less warm surface water and therefore abundant ocean life; the water taken from the oceans to form the great glaciers would have lowered the sea level by 120 metres and this would have provided an area of land as large as Africa on which plants could grow. As we have seen, there was more life on the colder Earth, shown by the low abundance of carbon dioxide at that time; it takes a lot of life to pump it down to less than 200 parts per million (ppm). More than this, the ice-core evidence from Antarctica suggests that the output of dimethyl sulphide (DMS) was nearly five times greater in the ice age. This larger production of sulphur gas implies more marine algae, the source of DMS, in the oceans. In my view, if the Earth system, Gaia, could express a preference it would be for the cold of an ice age, not for today's comparative warmth.

There is much more to Gaia than temperature regulation. The maintenance of a stable chemical composition is similarly vital. Andrew Watson and Tim Lenton have gone far towards discovering the mechanism by which atmospheric oxygen is regulated and the part played by that important but rare element phosphorus. Peter Liss has investigated the biological sources in the oceans of the essential elements sulphur, selenium and iodine. The intricate links between algae living in the oceans, sulphur gas production, atmospheric chemistry, cloud physics and climate are slowly being uncovered in dozens of laboratories around the world. Now that Gaian regulation is accepted, even if not understood, there is a worldwide effort to uncover the Earth's vital statistics. Much of the detail is available in the book *The Earth System* by Kump, Kasting and Crane. It is well worth reading as a source, even if it is, in the American way, not as Gaian as it could be.

In 1994 one of the authors, my friend the American geochemist Lee Kump, and I published a paper in *Nature* that described a computer model of the Earth like Daisyworld but more realistic; instead of daisies, we had ocean algal ecosystems that affected climate by pumping down carbon dioxide and also by making white reflecting clouds. On the land masses we had forest ecosystems that also pumped down carbon dioxide and made clouds. The defining part of our model was the growth rate of organisms at different temperatures. We took the generally accepted values of the growth rates of algae and forest trees under ideal conditions where water and nutrients were unlimited. This data revealed that growth was best near 30°C and stopped below 0°C and above 45°C. We then took into account the real world constraints set by the physical properties of water. For the algae in the ocean the best temperature for growth would be close to 10°C, because above this the stable surface layer forms and shuts off the supply of nutrients. Similarly, on the land the upper limit of tree growth would be set by the rate of evaporation of water, and the optimum for trees was close to 20°C.

When we ran our model by either steadily increasing the input of heat from the sun or by keeping the sun constant but increasing the input of carbon dioxide, as we are now doing in the real world, the model showed good regulation, with both the ocean and land ecosystems playing their part. But as the carbon dioxide abundance approached 500 ppm, regulation began to fail and there was a sudden upward jump in temperature. The cause was the failure of the ocean ecosystem. As the world grew warm, the algae were denied nutrients by the expanding warm surface of the oceans, until eventually they became extinct. As the area of ocean covered by algae grew smaller, their cooling effect diminished and the temperature surged upwards.

Figure 1 shows a run of this model with a steadily increasing input of CO_2 pollution going from the pre-industrial level to up to three times as much, which is less than we are now adding to the atmosphere. The upper panel of the chart shows temperature change, with the upper line the temperature expected for a dead planet and the lower line for our model Earth. A feature of the model is a simple device to indicate if feedback is positive or negative. We introduced a small periodic variation in the heat received from the sun. The amplitude of this fluctuation was kept constant and is reflected in the variations of the otherwise constant temperature of the control dead planet shown in the upper line on the figure. The lower panel of the chart shows the changes in the land vegetation, in the ocean algae and in the carbon dioxide abundance. When regulation was working well, the abund-

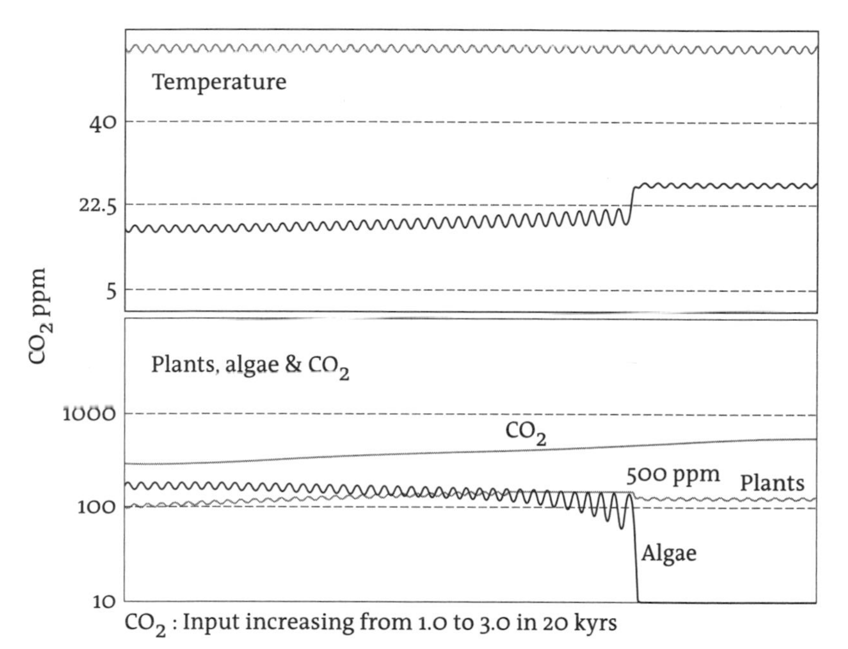

Figure 1. *Climate prediction according to the model described in the text.*

ance of the algae and plants and the temperature all show dampened fluctuations, but when the algal ecosystem became stressed the fluctuations grew large and showed amplification by positive feedback. The sudden jump in mean temperature from about 16 to 24°C followed the largest fluctuation and the extinction of the algae.

The model maps surprisingly well onto the observed and the predicted behaviour of the Earth. The turning point, 500 ppm of carbon dioxide, would, according to the IPCC, represent a temperature rise of about 3°C. This is close to the temperature rise of 2.7°C predicted by the climate modeller Jonathon Gregory as sufficient to start the irreversible melting of Greenland's ice. Those respected professional scientists who monitor the oceans and atmosphere already report an acceleration of the rise of carbon dioxide abundance and a decline in algae in the Atlantic and Pacific oceans as they warm.

I acknowledge that arguments from models like this one and from geophysiology are not by themselves strong enough to justify political action, but they become serious when taken in conjunction with the evidence from the Earth that nearly all the systems known to affect climate are now in positive feedback. Any addition of heat from any source will be amplified, not resisted, as would be expected on a healthy Earth. Of course, if we could manage to establish a net cooling trend the same positive feedback would work in our favour and accelerate cooling.

Some of these positive feedbacks are:

1) The ice albedo feedback first proposed by the Russian geophysicist M. I. Budyko ('albedo' refers to the reflectivity of an object or a surface). Ground covered by snow reflects almost all sunlight falling on it back into space and therefore stays cold. But once the snow at the edges begins to melt, dark ground emerges which absorbs sunlight and therefore gets warmer. Its warmth melts more snow, and with positive feedback melting accelerates until all the snow is gone. When the net trend is towards cooling, the same process operates in reverse. Just now the floating ice of the polar basin is rapidly melting and is an example of the Budyko effect in operation.
2) As the oceans warm, so the area covered by nutrient-poor water increases, making the ocean less friendly for algae. This reduces the rate of pump down of carbon dioxide and the generation of white reflecting marine stratus clouds.
3) On the land, increasing temperature tends to destabilize tropical forests and lessen the area they cover. The land that replaces the forest lacks cooling mechanisms and is hotter, and so, like the snow, the forest melts away.
4) Richard Betts, in a 1999 *Nature* paper, first observed that the Boreal forests in Siberia and Canada are dark and heat absorbing. As the world grows warmer they extend their range and so absorb more heat.
5) As forest and algal ecosystems die their decomposition releases carbon dioxide and methane into the air. In a warming world this also acts as a positive feedback.
6) Large deposits of methane are held in ice crystals within molecular-sized voids, called clathrates. These are stable only in the cold or

under high pressure. As the Earth warms there is an increasing risk of these clathrates melting, with the escape of large volumes of methane, which is twenty-four times as potent a greenhouse gas as carbon dioxide.

There are almost certainly other systems, both geophysical and geophysiological, that affect climate that we have not so far discovered, but the rate of global warming suggests that there is no large negative feedback that would countervail temperature rise. The only system we do know of that acts in negative feedback† is the long-term weathering sink for carbon dioxide, called 'rock weathering'.† This is the biochemical process by which carbon dioxide dissolved in rain water reacts with calcium silicate rocks. Vegetation on the rocks greatly enhances the removal of carbon dioxide, and the greater warmth leads to faster vegetation growth, making a stronger sink for carbon dioxide. But too much heat on the land masses could turn this also to positive feedback. There is also a negative feedback caused by fierce tropical storms, which stir the water sufficiently to draw up nutrients from below the surface layer and so allow algal blooms. We do not yet know how large an effect this has on climate.

Past and present atmospheric pollution with carbon dioxide and methane is similar to the natural release of these gases fifty-five million years ago, when comparable quantities of carbon entered the atmosphere. Then the temperature rose about 8°C in the temperate northern regions and 5°C in the tropics; the consequences of this heating lasted 200,000 years.

THE NATURE OF REGULATION

Until recently we accepted that the evolution of organisms takes place according to Darwin's vision, and the evolution of the material world of rocks, air and ocean according to textbook geology. But Gaia theory sees these two previously separated evolutions as part of a single Earth history, where life and its physical environment evolve as a single entity. I find it helpful to think that what evolves are the niches, and organisms negotiate for their occupancy.

The ideas I have just presented are part of the basis of Gaia theory, but a full explanation would require an account of how self-regulation works. In some ways this is not just difficult, it is impossible: emergent phenomena like life, consciousness and Gaia resist explanation in the traditional cause-and-effect sequential language of science. Emergence has similarities with the quantum phenomena of 'entanglement', and we may never be fully able to explain them. What we can do is express them in the language of mathematics and use them in the cornucopia of our inventions. Engineers are well able to design complex self-regulating systems, such as automatic pilots for ships, aircraft and spacecraft; communications engineers and cryptologists are already making devices that exploit quantum entanglement. But I doubt if any of them have a conscious mental image of their inventions; they develop and understand them intuitively.

To recapitulate, the part of Gaia thinking that most confuses is the question: what is self-regulation? What first amazed me about the Earth system was its capacity to stay close to the right temperature and the right chemical composition for life and to have done so for over three billion years, a quarter of the time the universe is thought to have existed. But for many years after the intuition of Gaia, I had no idea how it worked.

When I was about ten years old I was taken by my mother and father on winter Sundays from our home in Brixton to South Kensington. Their destination was the Victoria and Albert Museum, filled with art treasures, and mine was the Science Museum. Like most boys of that time, 1928 to 1932, I was fascinated by mechanical things and wanted to know how they worked. One of the exhibits was a working model of the steam engine, complete with James Watt's famous governor. This device regulates the engine's speed, and it consists of a vertical shaft driven by the engine on which is mounted two arms that carry iron balls at their ends. The arms are hinged to the shaft so that, as the shaft rotates, the balls swing out. The faster the engine runs, the higher the balls are lifted; a second pair of arms connected to those carrying the rotating balls simply lifts a lever controlling the flow of steam from the boiler of the engine. The faster the engine runs the more the steam valve is closed. It was obvious to me as a child that the engine would settle down to run at a constant speed, and that simply by changing the setting of the connection to the steam valve the speed could be set as high or as low as one wished. This was an early example of a control system using a negative feedback to govern the otherwise uncontrollable engine. Without it, the machine would race and perhaps shake itself to pieces when the steam pressure was high, or stop or run too slowly when the pressure was low. But was it really this simple?

James Clerk Maxwell was arguably the greatest physicist of the nineteenth century; in his mind the forces of magnetism and electricity were brought together in a comprehensive electromagnetic theory, a theory that laid the foundations of much of modern physics. Maxwell is reported to have said, a few days after seeing Watt's spinning ball governor, 'It is a fine invention, but try as I may, its analysis defies me.' Maxwell's puzzlement was not so surprising. Simple working regulators, the physiological systems in our bodies that regulate our temperature, blood pressure and chemical composition, and simple models like Daisyworld, are all outside the sharply defined boundary of Cartesian cause-and-effect thinking. Whenever an engineer like Watt 'closes the loop' linking the parts of his regulator and sets the engine running, there is no linear way to explain its working. The logic becomes circular; more importantly, the whole thing has become more than the sum of its parts. From the collection of elements now in operation a new property, self-regulation, emerges – a property shared by all living things, mechanisms like thermostats, automatic pilots, and the Earth itself.

The philosopher Mary Midgley in her pellucid writing reminds us that the twentieth century was the time when Cartesian science triumphed. It was a period of excessive hubris and called itself the century of certainty; at its start there were eminent physicists saying, 'there are only three things left to discover', and at the end they were seeking the 'theory of everything'. Now in the twenty-first century we are beginning to take seriously the remark of that truly great physicist, Richard Feynman, about quantum theory: 'anyone who thinks they understand it probably does not.' The universe is a much more intricate place than we can imagine. I often think our conscious minds will never encompass more than a tiny fraction of it all and that our comprehension of the Earth is no better than an eel's comprehension of the ocean in which it swims. Life, the universe, consciousness, and even simpler things like riding a bicycle, are inexplicable in words. We are only just beginning to tackle these emergent phenomena, and in Gaia they are as difficult as the near magic of the quantum physics of entanglement. But this does not deny their existence.

3

The Life History of Gaia

Life on Earth began between three and four billion years ago; we can only guess the date, since there are so few unequivocally dated fossils to be found. At this early time the sun was probably 23 per cent less luminous than it is now. We think that the Earth was mainly covered by ocean and there were only small continents. It would have been kept warm enough for water to stay liquid and for life to start through the presence of abundant carbon dioxide in the atmosphere, perhaps thirty times more than now, and it may have been a darker planet than now, because there was less land and possibly fewer clouds. Once photosynthesis evolved it would have used the carbon dioxide as its carbon source and by so doing decrease its abundance in the air.

We could look at this as a reverse greenhouse effect that presented early life with problems like the greenhouse warming that we face today, but for early life the threat was cooling or freezing, not warming. We think early life resolved this problem through the evolution of organisms called methanogens, which are still around in our guts and anywhere there is a lack of oxygen. These 'detritophores' live by decomposing the bodies of deceased photosynthesizers and other organisms; the main products of their decomposition are the gases methane and carbon dioxide. Methane is twenty-four times as potent a greenhouse gas as carbon dioxide, and when its atmospheric abundance was about 100 ppm in the early Earth's atmosphere it would easily have kept our infant planet warm enough for life. This idea, first mentioned in my book *The Ages of Gaia* in 1988, is slowly becoming the conventional wisdom among geochemists.

Once Gaia came into existence as a planetary system (and I think that this would have been some time after life itself had started) it would have changed the atmosphere from one dominated by carbon dioxide to one dominated by methane. This ancient world of bacteria would have been dynamically stable and resilient against perturbation, but the departure from the stable equilibrium state of a dead planet would have made Gaia vulnerable to catastrophes, such as planetisemal impacts or huge volcanic outbursts. If an event of this kind removed most of the living organisms, the methane would rapidly have vanished from the air and the Earth would have frozen; but in those early times recovery was automatic, as carbon dioxide vented into the air from volcanoes and built up a greenhouse that re-warmed the Earth. There would have been enough survivors to rebuild the smelly septic-tank world of our infant Gaia. Things are very different now; any catastrophe that caused Gaia's regulation system to fail would lead to a hot and dead Earth with no natural means of returning back to its cooler state.

Simple models of Gaia are stable and not easily perturbed, but only if more than a critical mass of life is present on the model planet. The models usually come to equilibrium with 70 to 80 per cent of the planetary surface inhabited, the remainder assumed to be barren or sparsely populated desert or ocean. If a plague or some other mishap destroys more than 70 to 90 per cent of the population, the temperature and chemical composition cease to be regulated and the model system swiftly drops to the equilibrium state of the dead planet.

The vulnerability of these model systems to upsets depends upon the intensity of the stress the planet is undergoing before the disturbance occurs. With a model of the Earth two billion years ago I found that almost all of the living organisms could be eliminated without disturbing the planetary climate. At this time the Earth was briefly passing through its 'Goldilocks' stage, when the heat from the sun was just right for life and little or no temperature regulation was needed. This may have been why one of the great crises of Gaia's existence, the appearance of oxygen as a dominant atmospheric gas, passed without deadly consequences. It happened when the climate of the solar system was benign. At the beginning, over three billion years ago, the sun was too cool for comfort – now it is too hot.

The appearance of oxygen was an event as important in Gaian history as puberty is in humans. It drove the evolution of more complex living cells, the eukaryotes and eventually the huge assemblies of living cells that make up plants and animals. Not least, it allowed the Earth to retain its oceans by acting as a barrier against the escape of hydrogen to space. For over a billion years after oxygen appeared, the evolution of life on Earth passed through something like a dark age, with little or no historical evidence. This period, the proterozoic, was one where life was still unicellular, and it left behind in the geological record almost nothing in fossil form.

Our view of the Earth's past is like that of a landscape from a mountain viewpoint. Apart from a few other snowy distant peaks, large forests and lakes, nothing detailed is discernible beyond a mile or so; the history of the British Isles in the ice ages of the Pleistocene falls in this discernible range. During the brief warm interglacials, it seems to have been an unbroken, shore-to-shore carpet of trees, a broad-leafed temperate forest ecosystem, small compared with the huge tropical rainforests of today, but like them diverse in its range of species. The carpet of trees covered nearly all of the land, including the mountain areas that are now treeless; indeed, what is often spoken of now as wilderness was then covered in trees. Grazing animals would have made a few clearings and forest paths, but these would represent only a tiny fraction of the whole. A bird flying high over the British Isles would have seen a densely packed forest extending to the horizon, just like a present-day aerial photograph of Amazonia.

I find it remarkable that such a verdant scene has alternated more than twenty times, with much longer periods of tundra and glaciers that, seen from above, would have looked like Greenland today. The long ice ages swept away the trees and all but sterilized the land; yet when the climate warmed for the short interglacials, life returned anew and in much the same way every time. The frostbitten extremities of the Earth healed well when a warmer climate came.

As a geophysiologist, I look on these cold and warm events as a series of experiments. Trees and other plants were seeded onto the warm but sterile land that was set free as the glaciers retreated, and they rapidly grew until there was confluent forest cover. Then the experimental region was put in the deep freeze of a glaciation until it was time for a repeat. It was a good series of experiments, and in the many repetitions the results varied only by a small amount. A botanist, for example, would notice variations in the organisms present: sometimes there would be mainly oak, while in other, colder periods, alders, birch and conifers would predominate.

I suspect, but do not know, that the biodiversity – that is, the number of different species present in a defined area – would also have changed. Stable unchanging climates lasting for several thousand years tend to reduce diversity, but when the climate changes to either hotter or colder by a small amount the first response is an increase in biodiversity. This is because the new conditions give rare species a chance to flourish while the established ones have not had time to decline. When the climate stabilizes again, survivors of the past regime may die out and biodiversity diminish again. Of course, biodiversity falls almost to zero in the impoverished environment of the glaciation, but it is important to keep in mind that biodiversity and environmental quality are not simply proportional.

A planetary physician would look on biodiversity as a symptom, a response to change. He would recognize that what is a rare species in one state becomes a common one in another. So rich biodiversity is not necessarily something highly desirable and to be preserved at all costs. A red, flushed and sweaty skin is our physiological response to overheating, and the biodiversity of a tropical forest like Amazonia may be the Earth's response to the heat of the present interglacial. Neither of these states is worth preserving as a long-term goal, and evolution would change them into something more stable. I suspect that the capacity to become biodiverse has evolved because, in the real world of Gaia, change is always happening and is usually driven from outside by small alterations in the seating arrangements of the solar system and in the output from the sun. When there is a climate change, dormant seeds, rare plants, or seeds drifting in on the wind, or on the feet of birds, have a better or worse chance to grow; if better, they flourish and compete with the native species until they become a stable part of the ecosystem. During the period of competition biodiversity is increased, but it declines again as the ecosystem adapts to the new conditions.

We have become so concerned over the fate of the rare tree, especially if it produces a drug that might cure cancer, and about rare and beautiful animals and birds; we have become so excited by these collectables that we have lost sight of the forest itself. But Gaia's automatic response to adverse change is driven by the changes in the whole forest ecosystem, not by the presence or absence of rare species alone. Niches vacated by extinction do not stay empty, and like the great rentier that she is, they are rapidly occupied; her rent, the cash flow of elements, is just as well paid by dull and abundant plants as it is by rarities – as with the human ecosystem of London, which displays its exotica in the habitats of Hampstead, Notting Hill, and Islington.

But what of the glaciations, when it grows really cold and ice begins to scrape away the soil and destroy almost all life? Why does Gaia not resist this adverse change? The answer lies, I think, in a long-term, whole planet view. As the aeons have passed, the sun has remorselessly grown hotter; that is the nature of the nuclear furnaces that power stars, and as they age they increase their heat output and eventually die in a burst of fire. In order to sustain an equable climate the Earth system has evolved several air-conditioning mechanisms. Vegetation growing on the land and floating in the sea uses carbon dioxide that it removes from the air, and this lessens the carbon dioxide abundance and its greenhouse effect; another mechanism is the production by marine organisms of gases that, when oxidized in the air, make the tiny particles called cloud condensation nuclei, without which water in the air would not condense as the droplets that clouds are made of. Without clouds, the Earth would be much hotter.

The period we are now in is close to a crisis point for Gaia. The sun is now too hot for comfort, but most of the time the system has managed to pump down carbon dioxide sufficiently and to produce enough white reflecting ice and clouds to keep the Earth cool and to maximize the occupancy of the Earth's niches. But to do so the regions above 45° north and below 45° south of the equator have had to be sacrificed. This is not as large a loss to Gaia as it is to humans. These polar regions occupy less than 30 per cent of the Earth's surface, and their white reflecting surfaces powerfully assist cooling.

During an ice age, so much water is locked up in the glaciers of the polar regions that the sea level drops by 120 metres. Consequently, a vast area of land emerges from the sea, and much of it is in the tropics; Tim Lenton reminded me that the land released by the fall of sea level was equal in area to that covered by ice. The loss of productivity in the temperate and polar latitudes is more than compensated for by the increase in land life in the tropics and in the cooler oceans. Although there is a smaller area of ocean in an ice age, it is more productive, because cold water favours the growth of the primary producers, the photosynthetic algae. As I mentioned earlier, a warm ocean is, perversely, nowhere near as productive as a chilly one. The colder waters are the dense forests of the sea, rich in life and helping to keep the Earth cool by producing clouds and by pumping down carbon dioxide.

THE SENESCENCE AND DEATH OF GAIA

The energy source of the solar system is the sun. This nuclear furnace has now operated for four and a half billion years and will continue for about another five billion, when its supply of fuel – hydrogen and helium – runs out. In the long term the sun is not renewable, but in our terms it can be taken as so. The sun is a remarkably steady and reliable source of light and heat, and the supply is 1.35 kilowatts of energy for every square metre of the Earth that is in direct unimpeded sunlight.

Because the sun grows hotter, the heat received by the Earth now is more than it was when life began over three billion years ago. Yet most textbooks and television programmes on science will tell you that the Earth, like Goldilocks, is a planet that happened to be born at exactly the right distance from the sun, and this is why conditions on Earth are exactly right for life. This pre-Gaia statement is wrong, and only for a brief period in the Earth's history was the sun's warmth ideal for life, and that was about two billion years ago. Before this it was too cold for comfort and afterwards it has progressively grown too hot. In the very long term, solar warming is a far greater problem for life than our present-day battle with man-made global heating.

In about one billion years, and long before the sun's life ends, the heat received by the Earth will be more than two kilowatts per square metre, which is more than the Gaia we know can stand; she will die from overheating. Gaia regulates its temperature at what is near optimal for whatever life happens to be inhabiting it. But, like many regulating systems with a goal, it tends to overshoot and stray to the opposite side of its forcing. If the sun's heat is too little the Earth tends to be warmer than ideal; if too much heat comes from the sun, as now, it regulates on the cold side of ideal. This is why the usual state of the Earth at present is an ice age. The recent crop of glaciations the geologists call the Pleistocene is, I think, a last desperate effort by the Earth system to meet the needs of its present life forms. The sun is already too hot for comfort. The low level of carbon dioxide gives a measure of the problems faced by Gaia during an ice age; planetary life pumps down carbon dioxide from the air until it reaches levels as low as 180 ppm. This is half of what is in the air now and is too little for some plants to grow well. Michael Whitfield and I calculated, in 1981, that in less than 100 million years the sun's heat will be too much for the Earth to regulate at its current state, and it will be forced to move to a new hot state inhabited by a different biosphere. The brief interglacials, like now, are, I think, examples of temporary failures of ice-age regulation. These ideas were taken up and extended by Jim Kasting and Ken Caldiera in 1992 and by Tim Lenton and Werner von Bloh in 2001.

Looked at on this long-term and large scale we sense that our adding carbon dioxide to the air and soon doubling its abundance is seriously destabilizing an Earth system already struggling to maintain the desired temperature. By adding greenhouse gases to the air and by replacing natural ecosystems, like forests, with farmland we are hitting the Earth with a 'double whammy'. We are interfering with temperature regulation by turning up the heat and then simultaneously removing the natural systems that help to regulate it. What we are now doing is uncannily like the series of foolish actions that led to the Chernobyl nuclear reactor accident. There the engineers turned up the heat after they had disabled the safety systems, and it should have been no surprise that the reactor ran into rapid overheating and caught fire.

Climatologists now think that we are perilously close to the threshold beyond which adverse change sets in; change that is, on a human timescale, irreversible. The Earth does not catch fire, but it becomes hot enough to melt most of the Greenland ice and some of the West Antarctica ice; enough water will then be added to the world oceans to raise sea levels by fourteen metres. It is sobering to think that nearly all of the present great centres of population are currently below what could be the ocean surface in a mere blink of geological time.

It would be wrong to leave this account of Gaia without touching again on the fact that she is old and has not very long to live. As the sun grows ever hotter it will, in Gaia's terms, soon become too hot for animals and plants and many of the microbial forms of life. I think it unlikely that heat-tolerant bacteria, thermophiles living in the oases of a desert world, would be abundant enough to form the critical mass of living things needed for Gaia. It is also unlikely that the kind of Earth we know now would last even a fraction of those billion years. The harm done by a planetesimal impact, or even by a future industrial civilization, may drive Gaia first to one of the hotter and temporarily stable states, and finally to total failure.

Growing old is not as bad as is sometimes imagined. When I was in my teenage years it seemed then that by now I would be feeble, depressed and barely even half-witted. Some, but not all, of these premonitions have come true, and although I can walk and climb a modest slope at four miles an hour, walking at that speed over mountains is no longer an option. But somehow I learnt that life begins anew at each decade; it certainly, for me, began afresh at each decade from the age of 20 onwards. As with a butterfly, the long years as a grub and then a pupa are over, and as the poet Edna St Vincent Millay said:

> My candle burns at both ends;
> It will not last the night;
> But, ah, my foes, and oh, my friends –
> It gives a lovely light.

So it is with Gaia. The first aeons of her life were bacterial, and only in her equivalent of late middle age did the first meta-fauna and meta-zoa appear. Not until her eighties did the first intelligent animal appear on the planet. Whatever our faults, we surely have enlightened Gaia's seniority by letting her see herself from space as a whole planet while she was still beautiful. Unfortunately, we are a species with schizoid tendencies, and like an old lady who has to share her house with a growing and destructive group of teenagers, Gaia grows angry, and if they do not mend their ways she will evict them.

4

Forecasts for the Twenty-first Century

Michael Crichton argues that long-range weather prediction is impossible because of the chaotic[†] mathematics of weather systems. Most professional meteorologists would agree with him, but he is quite wrong when he says that the same is true of climate prediction.

Future climates are much more predictable than is future weather. We know that there is no way to predict if it will, or will not, rain on 2 November 2010 in Berlin. But we can with near certainty say that it will be colder in January in that city than it was in the previous July. Climate change is amenable to prediction, and this is why so many scientists are tolerably sure that a rise of carbon dioxide to 500 ppm, which is now almost inevitable, will be accompanied by profound climate change. Their confidence comes from knowledge of the past history of the many glacial and interglacial events of the past two million years. The record drawn from the analysis of Antarctic ice cores clearly shows a strong correlation between global temperature, carbon dioxide and methane abundance.

If any one of us wants to know the social conditions of Victorian England we go to Dickens, Trollope and the other fiction writers of that time. More than this, we speak about their writings as if they were the true historical account. This is why I take Michael Crichton's opinions seriously, not because they are true, but because he is such a good storyteller; indeed, he is among my favourite authors of a good yarn (his mix of medieval history and quantum theory in his book *Time Line*, for example, made it the best of science fiction). The public is much more likely to be influenced by writers like Michael Crichton than they are by scientists. Fiction writers and film producers should ask themselves if they are sure that what they say is true before succumbing to the overriding imperative of the storyline; this is more important than ever before, now that we face deadly change.

The authoritative source of information and prediction on the climate of the coming century is the Intergovernmental Panel on Climate Change (IPCC). The IPCC issued its third assessment report in 2001, and the next is due in 2007. Sir John Houghton, formerly the director of the UK Meteorological Office, was one of the joint chairmen of the IPCC, and his book *Global Warming*, with its third edition published in 2004, provides the most up-to-date and readable account of our understanding of this fast-changing field of science. It is revealing to look back at the climate forecasts made in the late 1980s. Here, from Stephen Schneider's 1989 book *Global Warming*, is a chart that illustrates the thoughts of climate scientists at a conference in 1987 (Figure 2). From the limited knowledge then available they did their best to predict the future climate and showed their guesses as dotted lines on the graph. The upper dotted line is of a scenario they thought almost science fictional in its extremity. The cross I have added to the chart shows where we are now: we are already close to the extreme temperature change that made those pioneers so anxious.

Future climate predictions are mostly based on mathematical models of the Earth that were first used to try to predict weather a day or so ahead. These weather models divided the whole atmosphere into small parcels and calculated separately and in combination the changes likely in each parcel. To do this fast and well needs a fairly powerful computer; interestingly, so advanced are home computers now that yours may be powerful enough for a modest model of this kind. When it comes to climate prediction it is not enough to consider just the physics of the atmosphere. We need to take into account the way that the ocean stores heat and carbon dioxide and the dynamics of its interchanges with the atmosphere; we also need to know the nature of the land surface – whether or not it is covered with snow makes a huge difference, for example. Forests we now know are not passive areas on a map with fixed climate properties but are live actors in the climate system; the same is true of the ocean surface and the organisms that live in it. The clouds and the dust particles suspended in the air

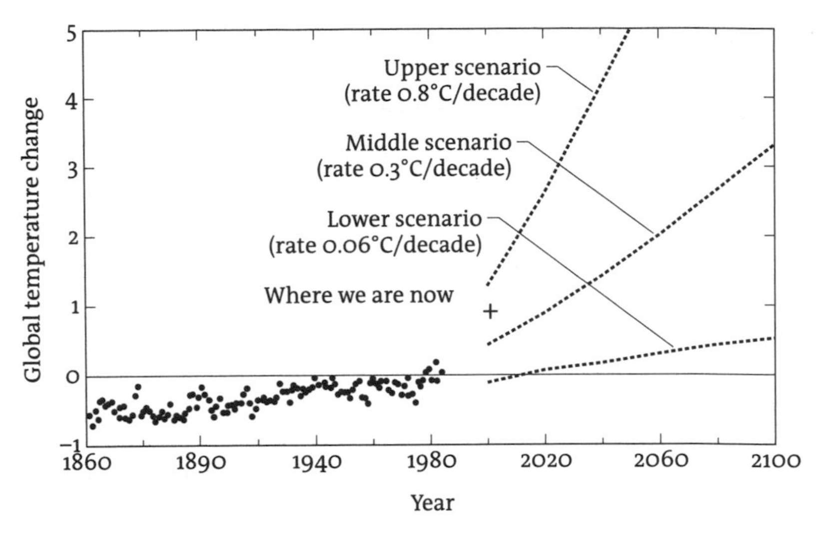

Figure 2. *Climate forecasts made in 1988.*

also have a powerful effect on climate. To take account of all the vast number of variables, we need a large computer. Fortunately, we have at the Hadley Centre in Exeter, UK, and in Japan, at their science city, Tsukuba, the largest climate models in the world, and scientists from the two institutions collaborate. But in spite of the expertise and the powerful computing machinery, our forecasts are provisional and do not include all surprises. Some, like the threshold of irreversible change, we think exist, and we wonder if the circulation of warm and cold water in the North Atlantic may be poised for sudden change. But we are not much better at dealing with the unexpected than were Columbus and his sailors when they set sail westwards for the East Indies. Their model of a round Earth was good, but the real planet had a huge and unpredicted surprise, the existence of the North American continent. We would be wise to expect that instead of temperature and sea level rising smoothly as the years go by, as in the IPCC predictions, there will be sudden and wholly unpredicted discontinuities.*

* Should you wish to enjoy some hands-on experience of modelling climates, there can be no better way to do it than through Kendall McGuffie and Ann Henderson-Sellers' book *A Climate Modelling Primer*, 2005. The book comes with a CD bearing programs of models that will run on most personal computers.

There are several reasons to think that our journey into the future will not be plain sailing and that one or more thresholds or tipping points do exist. Jonathon Gregory and his colleagues at Reading University reported in 2004 that if global temperatures rise by more than 2.7°C the Greenland glacier will no longer be stable and it will continue melting until most of it has gone, even if the temperatures fall below the threshold temperature. Because temperature and carbon dioxide abundance appear to be closely correlated, the threshold can be expressed in terms of either of these quantities. The Hadley Centre scientists Richard Betts and Peter Cox conclude that a rise in temperature globally of 4°C is enough to destabilize the tropical rain forests and cause them, like the Greenland ice, to melt away and be replaced by scrub or desert. Once this happens the Earth loses another cooling mechanism, and the rate of temperature rise accelerates again. In Chapter 1 I describe a simple model where the sensitive part of the Earth system is the ocean; as it warms, so the area of sea that can support the growth of algae grows smaller as it is driven ever closer to the poles, until algal growth ceases. The discontinuity comes because algae in the ocean both pump down carbon dioxide and produce clouds. (Algae floating in the ocean actively remove carbon dioxide from the air and use it for growth; we call the process 'pumping down' to distinguish it from the passive and reversible removal of carbon dioxide as it dissolves in rain or sea water.) The threshold for the failure of the algae is about 500 parts per million (ppm) of carbon dioxide, about the same as it is for Greenland's unstoppable melting. At our present rates of growth we will reach 500 ppm in about forty years. The monitoring now in progress of all these crucial parts of the Earth system – Greenland, Antarctica, the Amazon forests and the Atlantic and Pacific oceans – shows a trend towards what on our timescale could be irreversible and deadly change. Indeed, the science editor of the *Independent* newspaper, Steve Connor, reported on 16 September 2005 the statements of several climatologists who had found the melting of Arctic ice to be so rapid that we may already have passed a tipping point.

Deadly it may be, but when we pass the threshold of climate change there may be nothing perceptible to mark this crucial step, nothing to warn that there is no returning. It is somewhat like the descriptions some physicists give of the imagined experience of an astronaut unlucky enough to fall into a massive black hole. The threshold of no return from a black hole is called the event horizon; once this distance from the centre of the hole is passed gravity is so strong not even light can escape. The remarkable thing is that the astronaut passing through would be unaware; there is no rite of passage for those passing thresholds or event horizons.

For several years now I have had on the wall above my desk that amazing graph of the temperature of the northern hemisphere from the year 1000 to the year 2000. It was produced by the American scientist Michael Mann from a mass of data from tree rings, ice cores and coral. Part of the version in the 2001 IPCC report is reproduced below. It is called in America, mostly by sceptics, the 'hockey stick' graph. This is because it looks like a hockey stick lying flat with its striking end pointing upwards. I keep it in view to reinforce my arguments with sceptics of global heating and also as a reminder of how severe it will be. The graph shows the natural fluctuations of tempera-

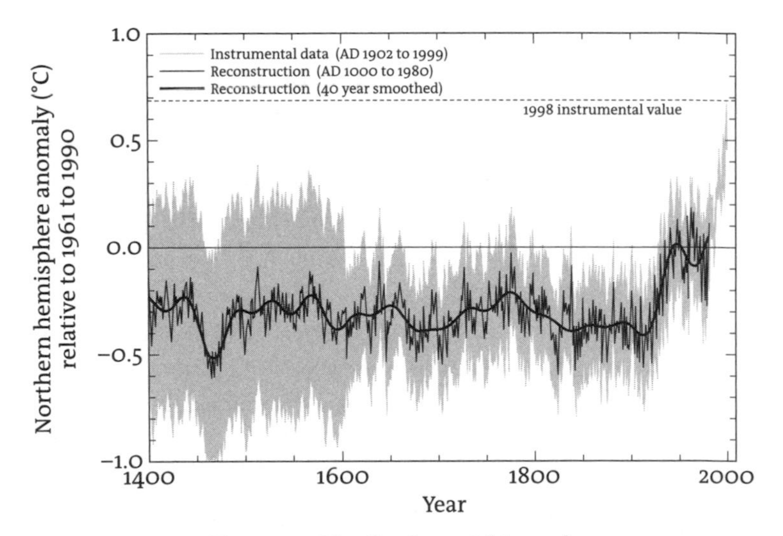

Figure 3. *The 'hockey stick' graph.*

ture, and for the first 800 years of the past millennium there is a slight but perceptible downward trend, which, if projected, points to an ice age in about 10,000 years. Then, at the start of the industrial period in about 1850, it slowly begins to rise, and with ever-increasing acceleration it climbs to reach temperatures nearly 1°C above the long-term average. A single degree rise in temperature may seem trivial, but remember we are looking at an average for half the world, the northern hemisphere. The difference between the long-term average of the graph and the ice age, 12,000 years ago, is just over 3°C. The IPCC 2001 report suggests that the line of the hockey stick graph might rise a further 5°C during this century. This is about twice as much as the temperature change from the ice age to pre-industrial times.

Every 25,000 years or so, the position and inclination of the Earth to the sun changes so that there is a small increase in the total flux of warmth the Earth receives. On every third of these successive pulses of extra heat Gaia has reached its lowest temperature and lowest carbon dioxide abundance; this is a sensitive state where the extra heat is more than can be managed and regulation fails. Gaia then enters an unstable state called an interglacial, much like a fever in one of us. This is the state of the Earth now.

We either forget or never knew how different the climate was in the last ice age. Most of the United Kingdom, and north-western Europe including Scandinavia, was buried beneath 3,000 metres of ice, a glacier as thick as that on Greenland now. North America was similarly glaciated as far south as St Louis, latitude 35°N. Despite all this ice it was probably a healthier world than now and more vegetation grew, both on land and in the sea. We think this because the abundance of carbon dioxide in the air was then below 200 parts per million. It takes a lot of life to pump it down that low.

The sea level was 120 metres lower than now, and land equal in area to the continent of Africa which is now below water was then above it. Much of this extra land was in South East Asia, which may explain why Australia was reached by humans during the ice age: the distance was short enough to be made on rafts or simple boats. Imagine there was a civilization 12,000 years ago with cities on the coast of that extended southern Asian continent. Who among them would have believed an early climate forecaster who claimed that soon they would be 120 metres beneath an ocean?

The changes likely in the world to come will, in their different ways, be as great as or greater than this. True, the sea cannot rise more than another eighty metres, the amount of extra water which would be released if the ice of Greenland and Antarctica melted. But the world-wide torrid conditions would reduce the productivity of the remaining land and sea, and the loss of vegetation would slow the rate of removal of carbon dioxide and so sustain the hotter age for 100,000 years or more. The greatest observable changes so far are in the Arctic, as was predicted in the first IPCC report in 1990. Below are satellite views of the Arctic basin in 1987, 2003 and an estimate view for some time between 2030 and 2050.

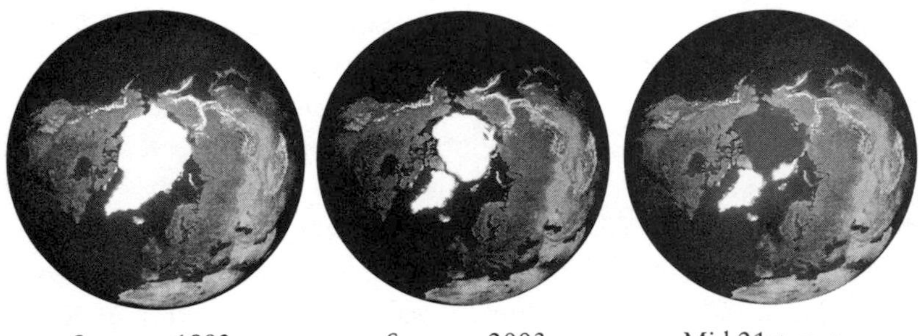

Figure 4. *The progressive summertime decline in the area of floating ice.*

The floating ice of the Arctic covers an area equal to that of the United States and serves as the home of polar bears and other animals; it is also the destination of the brave explorers who travel on foot to the North Pole. But, much more than this, it serves us all as a white reflector of the summer sunlight that falls upon it and helps to keep the world cool. When that ice melts, as soon it may, you will be able to reach the North Pole in a sailing boat, but we will have lost the air-conditioning capacity of the Arctic ice; the dark sea that replaces it will absorb the sun's heat and, as it warms, accelerate the melting of the Greenland ice.

While Gaia may suffer from the unfreezing of the Arctic basin and Greenland, these areas may become the future centres of an appropriately diminished civilization, and already shipping companies are beginning to prospect new polar routes. The Northwest passage that for so long has been barred by ice will soon open; the tundra wastelands of Siberia and northern Canada that remain above sea level will be rich with vegetation, and the enlarged Arctic Ocean, filled with algae, may become the fishing grounds of the future.

Another likely change often discussed by climatologists concerns the path of the great ocean conveyor belt that moves the waters of the world's oceans. The distinguished American Earth scientist, Wally Broecker, first warned us that the North Atlantic part of this conveyor depended upon the presence of Arctic conditions near Greenland. The waters that flow north on the surface of the Atlantic are warm and lose water by evaporation and so become saltier; salt water is denser than fresh water and it would sink were it not that the cold waters beneath are denser still. When this warm dense salt water is cooled by contact with the Arctic ice it sinks to the bottom of the ocean; the sinking provides the force that drives the conveyor and keeps moving the warmer salt water that drifts north-eastwards across the Atlantic, what we call the Gulf Stream. Broecker warned that if the down flow of cooled salt water ceased, northern Europe would no longer receive the benefit of this flow of warm water. Sensational fiction often portrays this as the return of Arctic conditions to northern Europe and the east coast of North America. But of course by the time it happens the Arctic ice will be well on the way to disappearance. I can't help wondering if the climate of the British Isles and the western part of northern Europe, which is now 8°C warmer than the same latitudes in other parts of the world, may be largely unchanged by global heating, because the 8°C lost when the Gulf Stream fails is just about equal to the predicted rise of temperature from global heating. Perhaps this is no more than wishful thinking, and we will certainly have to pay through the loss of land as the ocean rises to repossess it.

When we talk of climate change we often think more about the temperature and less about changes in the other qualities of the physical environment. Kangsheng Wu has pointed his research to the fresh water balance of the world and reported a persistent increase in the flux of fresh water to the oceans particularly in the North Polar basin. The freshening of these northern waters might alter the course of the Gulf Stream. In a similar way, increasing warmth may expand the Hadley cells (see pp. 80–81) and so cause a migration of the trade winds and the westerly winds to zones nearer the poles. Changes in these other properties of climate will surely happen as the Earth heats. The planners of large schemes for renewable energy using wind and water power need to keep in mind the likelihood that they may become expensive mistakes.

While we cannot go back to the achingly beautiful world of 1800, when there were only one billion of us, we may not be incapable of lessening the consequences of global heating. If there is a threshold and we pass it, the nations of the world could limit the damage by stopping carbon dioxide and methane emissions; the temperature rise would then be slower, as would the rise of sea level, and it would take longer to reach the final steady hot state than it would if we continued business as usual. Even so, enormous damage would still have been done. In a later chapter I will discuss proposals to use either terrestrial or space-mounted sunshades to cool ourselves back to pre-industrial temperatures. But even if we succeeded we would find ourselves saddled with the appalling responsibility of managing the Earth's climate, something that previously was provided free by Gaia, and we would still need to remove carbon dioxide from the air to prevent the poisoning of contemporary ocean life.

Recently the BBC broadcast in their *Horizon* series of science programmes an account of 'global dimming'; in it climate scientists, among them V. Ramanathan and Peter Cox, voiced their concern that we have already, in a sense, passed the point of no return in global heating. The science behind this programme appeared in a *Nature* article in 2005 which included as an author the distinguished German scientist, M. O. Andreae. Industrial civilization has released into the atmosphere, in addition to greenhouse gases, a huge quantity of aerosol particles, and these tiny floating motes reflect incoming sunlight back to space and cause global cooling. On large areas of the Earth's surface the aerosol haze reflects sunlight back to space sufficiently to offset global warming. By themselves they cause a global cooling of 2 to 3°C. Back in the 1960s, when we knew much less about the Earth and its atmosphere, a few scientists even speculated that continued economic growth would increase the density of the aerosol and lead to global cooling and even precipitate the next glaciation.

The present extent of aerosol cooling is real and seriously worrying. It may have allowed us to continue our business as usual, not noticing how much we had changed the Earth nor realizing that we would have to pay back the borrowed time. Aerosol particles stay only a brief time in the atmosphere: within weeks they settle to the ground. This means that any large economic downturn, or a planned reduction in fossil-fuel usage, or unwise legislation to stop sulphur emissions, as the Europeans are now enacting to stop acid rain, will allow the immediate expression of greenhouse warming. It has been suggested that part of the excessive heat of the 2003 summer in Europe was caused by the European Union's efforts to remove the aerosol which is the source of acid rain. Peter Cox reminded us that because the aerosol was not fully included, climate modellers may have underestimated the sensitivity of their models to greenhouse gas abundance and failed to notice that we may already be beyond the point of no return.

Predictions of climate change do not depend only on theoretical models in the form of computer simulations of the Earth. There is now a vast array of monitoring activities sustained globally. Air and sea temperatures are continuously measured, as are the gases of the atmosphere, the cloud cover, the floating ice and the glaciers and the health of the ecosystems in the ocean and on the land. The truth of the models is therefore continuously tested against the observations coming in from the real world. Satellites orbiting the Earth monitor its ever-changing scene. The more subtle instruments aboard these spacecraft monitor temperatures at different levels in the air and many different atmospheric gases; they also check the health of ecosystems. I often think that the great unsung wonder of the space programme is the way it has revealed so much about the Earth.

Another important source of information about the cause of climate change is the long-term geological record. We have learnt an immense amount about the history of the climate and the composition of the Earth's atmosphere from the analysis of ice taken from the depths of the Greenland and Antarctic glaciers. The snow falling on the glaciers brings air with it in the spaces between the crystals. Each new snowfall buries its predecessor, and so the air becomes trapped in small sealed bubbles made of ice, so that there is a continuous record going back to snow falling one million years ago. The bubbles trapped in the ice of cores bored into the glaciers provide samples of past atmospheres, and from their analysis the composition of these past atmospheres is revealed. From this vast data bank we now have a record, not only of the principal gases, oxygen and nitrogen, but also of the trace gases, carbon dioxide and methane. Indirectly we can calculate the temperature of the Earth when the air was trapped, from the isotopic composition of the oxygen and hydrogen. There are also good ways for ascertaining the date of the air being analysed. In this great store of information we have evidence that gives confidence to our claim that temperature and carbon dioxide abundance are closely correlated. We know that in the depth of the last glaciation carbon dioxide fell to 180 ppm, rose to 280 ppm after the ice age ended, and has risen now to 380 ppm as a result of our pollution. Already we have made as large a change in the atmosphere as occurred between the ice ages and the interglacials. If it stays at 380 ppm we might expect a comparable rise in temperature, but more probably as we continue to pollute it will rise to 500 ppm or more.*

Going further back in time, there have been hot spells similar to the one we think is now due. The most recent occurred fifty-five million years ago at the beginning of the geological period called the Eocene and is the subject of several papers by Professor Harry Elderfield of Cambridge University. It was in some ways similar to our pollution of

* On a visit to the British Antarctic Survey's laboratory at Cambridge, my wife Sandy and I were privileged to see the cores of Antarctic ice in which were bubbles of ancient air and hear them crackle and pop as melting released their pressure. We should be proud to have the British Antarctic Survey, under the outstanding leadership of Professor Chris Rapley, as monitors of the pathology that is disabling Gaia.

the air now and was due to the release of between 0.3 and 3.0 terratons of fossil carbon (a terraton is a million million tons). The source of this huge emission of carbon gases is still under debate: it may have come from the deposits of methane (natural gas) held in an ice-crystal form called a clathrate, which lie on the ocean floor, or it may have been vented from rich carbonaceous deposits in the North Atlantic when heated by a subterranean volcano.*

For comparison, we have already released by fossil fuel combustion and agriculture about half a terraton of carbon, a quantity within the range estimated for the Eocene hot event. There are differences between the Eocene catastrophe and our present day pollutions; for example, in the Eocene it was mainly methane that entered the air, not carbon dioxide as now. Professor Elderfield uses the geological record to suggest that fifty-five million years ago the temperatures rose by about 8°C in temperate regions and 5°C in the tropics, and from a world that was somewhat warmer than now, with little in the way of polar ice; the disturbance lasted 200,000 years. The sudden release of methane at the beginning of the hot spell would have rapidly warmed the Earth by its strong infra-red absorption, but it would have oxidized in the air to carbon dioxide and water vapour, and it would have been the carbon dioxide that sustained the heat for so long a period. The removal of carbon dioxide from the air by its chemical reaction with calcium silicate in rocks is called by geologists 'chemical rock weathering'. It is slow and takes about 100,000 years to remove 63 per cent of the gas. We now know from Gaia theory that life on the land surface and in the soil actively accelerates rock weathering. The land and ocean surfaces during the hot spell of the Eocene were barren, and that may be why the increased carbon dioxide stayed in the air so long. In addition, the Earth stayed warm because other biological cooling mechanisms that operate on a healthy Earth were disabled during the hot period of the Eocene. If conditions now are equivalent to those of the Eocene emissions, we should be prepared for a hot spell as long as or longer than an ice age. Although the initial conditions

* An account of this hypothetical event appears in an article by the Norwegian scientist Henrik Svensen and colleagues (*Nature*, 5 June 2004).

of the Eocene event resemble those on Earth now, two important differences are that the sun is now about 0.5 per cent hotter than it was fifty-five million years ago, equivalent to about 0.5°C in global temperature; and we have changed about half the Earth's land surface from natural forest into farmland, scrub and desert and consequently reduced the capacity of the Earth to regulate itself. There are now, in addition to carbon dioxide and methane, several other greenhouse gases whose presence in the air adds to global heating; these include the CFCs, nitrous oxide and other products of agriculture and industry.

The Earth has recovered after fevers like this, and there are no grounds for thinking that what we are doing will destroy Gaia, but if we continue business as usual, our species may never again enjoy the lush and verdant world we had only a hundred years ago. What is most in danger is civilization; humans are tough enough for breeding pairs to survive, and Gaia is toughest of all. What we are doing weakens her but is unlikely to destroy her. She has survived numerous catastrophes in her three billion years or more of life.

In spite of the heat there will still be places on Earth that are pleasant enough by our standards; the survival of plants and animals through the Eocene confirms it. It is possible that the British Isles, with its oceanic site and high latitude, will be one of these refuges, although it will be more an archipelago than the two main islands it now is. But if these huge changes do occur it seems likely that few of the teeming billions now living will survive.

I think it necessary to repeat that the smoothly rising temperature of the IPCC's third report shows an estimated average change of the global climate, but what it does not show are unpredicted extremes, including flood events and storms of great severity. We should expect climate changes of a kind never even thought of, one-off events affecting no more than a region. The first of these was the unprecedented European heatwave of 2003 when over 30,000 died of hyperthermia. Swiss meteorologists put the odds against it as no more than an unusually hot spell, at 300,000 to one.

Slower, decadal fluctuations in climate also confound our predictions. In a July 2005 *Science* article, Reading University scientists

Rowan Sutton and Daniel Hodson reported decade-long warming and cooling trends during the twentieth century in the North Atlantic climate and noted that the excessive heat of Europe in 2003 was in one of these warming periods, as was a similar period of warmth in the 1960s and 1970s. The present warm period follows a cooler climate in the 1980s. Variations of this kind are superimposed on the uprising curve of global heating, and we need to guard against an over-interpretation of unexpected warmth and cold as evidence for or against global heating.

In between the sceptics of global heating and those, like me, who are concerned at the possibility of drastic change, are the conservative climatologists who acknowledge global heating but think it unlikely to be severe. Among them are Tom Wigley and G. A. Meehl and his colleagues, who both have articles in the March 2005 issue of *Science*. These are good and thoughtful papers that forecast a world that will slowly heat by about 2°C and in which sea levels will rise between 10 and 30 centimetres by 2100, and assume rather drastic reductions in emissions. I surely hope that they are right, but I persist in my gloomier view of the future. I do so because several important properties of the Earth system may not have been included in their calculations. These are:

1) The possibility of the disappearance of the present man-made northern hemispheric aerosol. Because of its short residence time, an economic downturn or any of a number of disasters could cause it to decrease in a few weeks, leaving the greenhouse intact.
2) They may be neglecting the extent to which the Earth system is in positive feedback. This would make the sensitivity of their models to increasing greenhouse gases less than they expected.
3) They may not have included the feedbacks from the natural forests and the ocean algal ecosystems. These can make the forest a source of, instead of a sink for, atmospheric carbon dioxide when increasing heat causes the vegetation to die back. The same effect occurs with ocean algae as the seas warm and lessen the rate of pump down of carbon dioxide.
4) It is all too often assumed that the vast changes to the land surface

made by agriculture and forestry have had little or no influence on the sensitivity and resilience of the Earth system. I think it probable that the replacement of natural ecosystems with farmland may have altered the dynamics of climate feedback.

In its existence the Earth has experienced many different climate regimes. Soon after life started, Gaia emerged as a regulatory system; we think this led to a profound change in atmospheric composition from one dominated by carbon dioxide to one dominated by methane, which lasted about a billion years until oxygen became the chemically dominant gas, low in abundance at first but ultimately rising to become the air we now breathe.

Because temperature is so important to living organisms, it strongly affects their distribution on the Earth. Photographs of the Earth from space taken to show only the distribution of chlorophyll, the green pigment that vegetable life uses to convert sunlight into organic matter, provide a good way to grasp the effect of temperature on the geographical distribution of life. Chlorophyll is an essential constituent of all the primary producers that use the energy of sunlight to make food from the raw chemicals of the ocean and atmosphere; the distribution of chlorophyll represents that of plants and algae. It also shows where the other forms of life are because they consume it for food directly or indirectly. Figure 5 shows three sketch maps drawn to compare the distributions of plant and ocean algal life on worlds five degrees cooler, as now, and five degrees hotter than now. The centre sketch illustrates how at present the Antarctic continent and much of the North Polar region are almost bereft of life; the greater part of the world's oceans are also quite barren except for regions close to continents and in the cool waters nearer to the Arctic and Antarctic. The hot and dry deserts of Africa, Asia, North America and Australia are also sparsely populated with life. Abundant life occurs where it is warm and wet on land and where it is quite cool, less then 12°C, in the ocean.

Compare this with the two imaginary sketches: the lower an Earth 5°C hotter than now, roughly that predicted by the IPCC for the end of this century, and the upper 5°C cooler than now, close to the

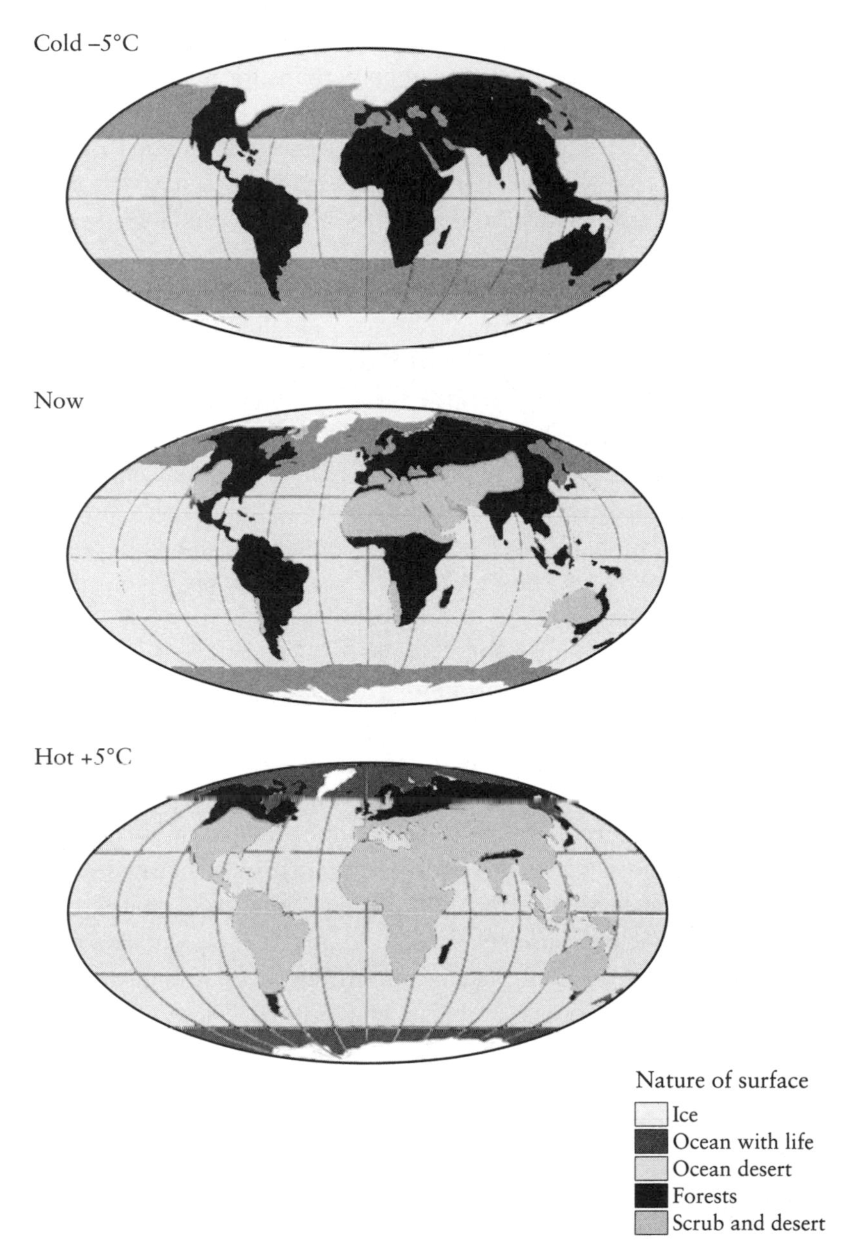

Figure 5. *The distribution of life now and on a hotter and a colder Earth.*

temperature of the last ice age. To judge from the abundance of life, Gaia seems to like it cold, which is why perhaps for most of the last two million years, and maybe much longer, the Earth has been in an ice age. I think it important that we recognize that a hot Earth is a weakened Earth. On the hot planet, ocean life is restricted to the continental edges, and the desert regions of the land are much extended.

What is remarkable is that life on Earth has persisted for nearly four billion years. This is a cosmic lifespan nearly a third as long as that of the universe itself. If life is so fussy about temperature then it implies that the Earth's temperature cannot have changed much during life's existence. We are tolerably sure that the sun, like all similar stars, warms up as it ages and is now 25 per cent hotter than when life began. This is equivalent to the temperature of the Earth's surface rising by 20°C; so that if the Earth was in an ice age when life began at 12°C it would by now be 32°C; or if it was warm, say 25°C, when life began, it might now be 45°C. Both of these are far above the 12°C of the glaciation and our present average of 16°C, and well above the 20°C expected from global warming.

This chapter has mostly dealt with climate change, but we should not forget that there are also large and disastrous changes in fresh water abundance, including floods and droughts. It is still far from clear whether the observed rise of sea level of the past fifty years is due mainly to the expansion of ocean water as it warms or mainly to the melting of glaciers. The 2001 IPCC report suggested that expansion was the principal cause of sea-level rise, but a 2004 *Nature* paper by scientists at the National Oceanic and Atmospheric Administration in Washington, and an essay by Jim Hansen in *Climate Change* in 2005, both suggest that the increase of ocean volume is mainly from the melting of land-based ice. If Hansen is right, then rapidly rising sea levels, not foul weather, will be the greater threat. The recent devastation wreaked by hurricanes in the Southern United States, especially in New Orleans, reminds us of the damage that temporary flooding can cause; excessive rain and surges of sea water driven by storms can be as disturbing as a permanent slow rise in sea level. All these changes in their turn alter the distribution of forests and deserts and the availability of land on which to grow food. Although we have much to learn, it does appear probable that in a few years, when the carbon dioxide abundance passes 500 ppm, we will enter the zone where temperatures will rise to a new steady state, perhaps six to eight degrees hotter than now. We do not know if this new regime will be stable in the long term, and if we are foolish enough to continue trying to farm and pollute the air on the remaining habitable parts of the Earth, a final collapse might happen. Nothing in science is certain, but Gaia theory is now robustly supported by evidence from the Earth and it suggests that we have little time left if we are to avoid the unpleasant changes it forecasts.

5

Sources of Energy

Lord Acton's dictum, 'Power tends to corrupt, and absolute power corrupts absolutely,' is usually associated with political power and is in danger of becoming a cliché, but it is also an alternative expression of the second law of thermodynamics, which says things tend to wear out, to run down and to become more disordered. It is not possible in this universe to use energy for any purpose, good or bad, without corrupting it.

The universe is said to have started in a primeval explosion of cosmic magnitude, the Big Bang, and perhaps this is why it is still almost wholly nuclear powered and a place where small planets like ours are lumps of radioactive fallout from the occasional star-sized nuclear explosions. Were it otherwise there would be no supply of the nuclear fuels, hydrogen, deuterium, uranium and thorium; nor would there be the internal heat of the Earth with its plate tectonics. Nuclear energy is not the most powerful source in the universe; gravity at the intensity exerted by black holes can convert matter into energy with an efficiency of near to 50 per cent, making it over 100 times more powerful than nuclear energy. Exploiting this energy source is at present the province of science fiction. In contrast to these abundant nuclear transactions of the universe, the photolysis of water by plants to make oxygen, and then the burning of stored carbon in it to gain energy, is one of the more bizarre biological phenomena of the solar system.

If it is perverse and dangerous to gain energy by burning fossil carbon in fossil oxygen, it may be equally so to imagine that comparable quantities of energy are freely and safely available from the so-called 'renewable' resources. Just imagine that we tried to power our present civilization on crops grown specifically for fuel, such as coppice woodland, fields of oilseed rape, and so on. These are the 'bio fuels', the much-applauded renewable energy source. Even if these natural products were used only for transport, to fuel our cars, trucks, trains, ships and aircraft, it would require us to burn every year about two to three gigatons of carbon as bio fuel (a gigaton is a thousand million tons). Compare this quantity with our yearly food consumption of half a gigaton; to grow this much already uses more of the Earth's land surface than may be safe. We would need the land area of several Earths just to grow the bio fuel. We may be foolish enough to go without food in order to drive as we wish, but Gaia is less tolerant. The land surface of the Earth has evolved as the site for ecosystems that serve the metabolism of the Earth, and they cannot be replaced by farmland. We have already taken more than half of the productive land to grow food and raw materials for ourselves. How can we expect Gaia to manage the Earth if we try to take the rest of the land for fuel production? What we can at least do is improve the efficiency of farming food by using a sensible proportion of farm and forestry waste, straw, manure and wood chippings as fuel.

Nor can we expect to gather all we need from wind, tide and solar energy without there being consequences. As good economists have warned, there are no free lunches, and already we are discovering that wind farms alter the vorticity of the atmosphere and may adversely change the climate in their vicinity. Their deployment so far is small. Will we cover Europe with wind farms only to discover harmful consequences we should have thought of before they were installed?

As I shall argue in more detail later in this chapter, I believe nuclear power is the only source of energy that will satisfy our demands and yet not be a hazard to Gaia and interfere with its capacity to sustain a comfortable climate and atmospheric composition. This is mainly because nuclear reactions are millions of times more energetic than chemical reactions. The most energy available from a chemical reaction such as burning carbon in oxygen is about nine kilowatt hours per kilogram. The nuclear fusion of hydrogen atoms to form helium gives several million times as much, and the energy from splitting uranium is greater still. This means that the amounts of nuclear fuel needed to supply our energy demands are tiny compared with Gaia's normal mass transactions, and so is the quantity of waste produced. We could use nuclear fission or fusion for quite some time before we ran into the kind of problem we are now having with fossil fuel.

It might seem that solar energy gathered from the sunlight falling on our rooftops is the ideal source. It might become so when efficient and cheap direct converters of sunlight to electricity are invented; as it is, they are still too costly for widespread use, although invaluable for powering spacecraft and remote monitoring equipment and wherever cost is not important. Heating water by sunlight is a sensible way of reducing the energy used from fossil fuel for domestic and industrial needs, and is widely used. There is one exception among renewable energy resources that is almost free of disadvantage, and that is geothermal energy. Unfortunately there are few places where it is freely available. Iceland is one of them, and it draws a large part of its energy needs from this source. But of course geothermal energy comes mainly from the heat generated by radioactive elements in the rocks and, like solar energy, is nuclear in origin.

When we talk of energy we tend mostly to think of electricity; at first sight this might seem blinkered, but as we shall see it is sensible to rank electricity first. It is true that a great deal of the energy we use comes from the direct combustion of fossil fuel; transport uses about a third of it and another main use is for home heating, while the rest goes to the manufacturers of steel, cement, plastics and all kinds of chemicals. Even so, a continuous uninterrupted supply of electrical energy is vitally important; it powers and sustains the nervous system of modern civilization. A twenty-first century city denied electricity would in a few weeks decline to a state comparable with a camp housing millions of starving and painfully uncomfortable refugees. Despite its importance, electricity is easily taken for granted – at least until the supply fails. There was a splendid drama-documentary shown on BBC television in March 2004 called *If the Lights Go Out*, which made clear the awful consequences of a national power failure later in this century. The drama was set at a time when supplies of natural gas came mainly from Russia and when 80 per cent of electricity came from power stations using gas as fuel. In the story, a terrorist incident ruptures the main supply line in Russia, leading to a total blackout of the UK. It vividly portrays the awful consequences for London of a complete cessation of electricity. No underground transport, no lights, including traffic lights, no fuel for cars, no lifts in high buildings, no heating (most systems are electrically controlled), no radio or television other than those run on batteries, and the failure of all central computers that had no alternative power supply. If all this were not bad enough, there would also be a widespread failure of food supplies, as refrigerators would no longer work, and water and sewage systems would probably fail. My insistence on the need for nuclear energy in this book comes because there is no other safe and reliable alternative for the large-scale production of electricity. The BBC drama was set at a time when a large anticyclone with calm cold weather made unavailable the tiny increment of electricity from wind farms.

To emphasize my contention that a continuous supply of electricity is an essential requisite for civilization, what follows is a brief account of our personal experience with a supply failure on a local scale caused by bad weather.

I was deep in thought before the screen of my word processor trying to answer an unusually difficult letter. A friend had written to seek my support. He was convinced that the power lines that ran near his home at Week St Mary, in Cornwall, threatened the health of his family; he was particularly fearful that the low-frequency radiation field had a potential to cause leukaemia. How could I say, not unkindly, that his worries were unfounded and that he had been led astray by the climate of fear that seems so much a part of life in our affluent first world?

Suddenly, my computer gave a strangled beep, the text on the screen faded and my room went dark. The storm that raged unnoticed by me in the double glazed quiet of my room had, in a burst of anger, brought down a tree across a feeder line that supplied the locality. The French have an odd but memorable euphemism for an orgasm: they call it 'la petite mort' – so powerful a feeling, yet not final like 'la grande mort'. It seemed perfectly to express my deep sense of dismay and loss that the failure of electricity always brings. How could my friend so fear electricity when I was distraught without it? To me it was instant cold turkey, that dreaded malaise that assails heroin addicts when the drug is withdrawn. Like it or not, I was surely, along with nearly everyone, mainlining on alternating current and had been for nearly all of my life.

Soon the house grew cold, and Sandy and I fitfully stumbled around trying to remember where we had put the butane gaslight after the last power failure. Several days passed before the electricity supply was restored; the storm was near hurricane force and was too much for the exposed lines of our rural supply.

I wondered how my friend and his family were faring: were their fears temporarily relieved or had the abrupt removal of the benefice of electric power changed their minds? There was no way to find out; the telephone lines were down and these were pre-mobile times.

We soon began to see that for us it was an unscheduled holiday, and we took time off to travel to the north Cornish coast near Morwenstow. Here the huge Atlantic swell bursts on the notoriously steep and dangerous rocky coast. We walked to the wooden hut built by the Reverend Hawker, the vicar of Morwenstow, one hundred years ago. We tried to imagine his thoughts as he watched for sailing ships to be dashed to pieces before his gaze.

More practically, we called on Theodore and Gerald who ran our local garage and we ordered a technological fix, a Honda generator to have as a standby. We were making the best of the misfortune but sorely missed the warmth of our home.

We were back to the older times; we had a home and shelter from the cold and wet, we had a wood-burning fire for warmth, and in one room, unlike the old days, we had a battery radio for news and entertainment. We hardly noticed that the television was unavailable; we rarely ever watched it anyway. The radio, especially BBC's Radio 3 and 4 programmes, could still inform and entertain, though we missed our hi-fi, since for us music is a key part of life. But most of all we had time to realize and talk about how utterly dependent even we who thought ourselves independent were on the constancy of electricity.

Electric power is extraordinary in the way it insinuates itself dendritically throughout society. The wires that carry it extend to every home like nerves in our bodies. It is always there at a wonderfully constant level, yet hardly anyone of the millions who work in the industry worldwide has more than the vaguest notion of what it is or how it is made and regulated. We are just like the termites who, without thought, build their spacious air-conditioned skyscraper communities. The electricity system exists as a communal, but almost wholly unconscious, activity. No wonder we took it for granted; at least we did until it stopped.

I find the most succinct and useful source of information about energy supply and use is Professor Michael Laughton's 2003 pamphlet *Power to the People*. It sails a steady course, undeflected by the special pleading of the energy producers or the green lobbies. What now follows is an account of the evolution of industrial and domestic energy sources seen from a planetary health, as well as from a human, perspective.

FOSSIL FUELS

From the time life began over three billion years ago, the dead remains of living things have been buried in the soil or in the muds at the bottom of the rivers, lakes and oceans. A small proportion, about 0.1 per cent, of the carbon in this organic waste escapes metabolism or fermentation by micro organisms and becomes part of the sedimentary rocks. Black coal and crude oil are tangible evidence of this bank of stored carbon, but the bulk of it is much more diluted and appears only as darker coloured sedimentary rocks. The other product of photosynthesis, oxygen, stays in the air. Before we began massively to interfere with this natural process, oxygen abundance was continuously regulated by the balance between the quantity of carbon buried and the rate of oxygen removal by reaction with carbon and other elements in rocks newly exposed by earth movement. The continuous movement of the Earth's tectonic plates, mainly driven by radioactive heat, piles up mountain ranges and exposes old buried sediments to weathering. As these are slowly worn away by rain, frost and ice, the buried carbon and other elements are exposed to the oxygen of the air with which they react, and the carbon becomes carbon dioxide again. The oxidation mainly occurs in micro organisms that gain energy by recombining the carbon and the oxygen. As used naturally in Gaia, fossil fuels provide wholly renewable energy, an inheritance from our ancestral life forms.

There is a naive belief that fossil fuels are unnatural and non-renewable. This false concept comes from seeing humans as supra-natural animals: fossil fuels are a product of living organisms and no more unnatural than a log of wood. When an accident at sea releases huge volumes of crude oil onto beaches, rocks and coves we see it as an environmental disaster, and not long ago we tried vainly to wash it away with detergents. Now, with greater common sense, we leave the clean-up to the natural organisms that regard the spillage as food.

When we burn fossil fuel for energy we are, in qualitative terms, doing nothing more wrong than burning wood. Our wrongdoing, if that is an appropriate term, is taking energy from Gaia hundreds of times faster than it is naturally made available. We are sinning in a quantitative not a qualitative way. Indeed, as I have already written earlier in this chapter, burning large quantities of wood or crops grown for fuel, something falsely considered as renewable energy, is potentially more destructive to the Earth system than using fossil fuels for energy. Both fossil and bio fuels are quantitatively non-renewable when burned at the excessive rate we require for our bloated, energy-intensive civilization. As always, we come back to the unavoidable fact that there are far too many of us living as we do now.

Coal and Oil

The generation of electricity by burning any available solid or liquid fuel has matured to a level of efficiency above which further improvements are unlikely. That old misery the second law makes it impossible for us to extract more than half the energy of burnt fuel as electricity. In reality, to get 40 per cent is good going; the other 60 per cent escapes as waste heat in the form of hot gases leaving the power station chimneys and steam rising from those strange and slightly menacing hyperbolic-sided cooling towers. (These are often associated with nuclear energy, but they are commonly used to improve the efficiency of any thermal power station.) In the more provident nations some effort is made to reduce inefficiency. Waste heat is collected as hot water and piped around the local town for winter heating. The extraction of carbon dioxide from the furnace gases is not a technically difficult problem and is already being done at a pilot plant in Norway. Here, the carbon dioxide sequestered is sent under pressure into a now-exhausted gas field under the Norwegian Sea. Sequestration and storage of recovered carbon dioxide will increase the cost of electricity from coal- and oil-burning power stations, but not beyond our means.

Chemical engineers have designed pilot plants that convert coal to hydrogen. The hydrogen can be burnt efficiently in a gas turbine, producing as waste only water vapour, and, when fuel cell technology matures, it can even more efficiently be converted directly into electricity. Carbon dioxide would still be a byproduct needing storage or burial, but these unconventional power stations appear to be efficient producers of electricity.

If only we had developed and installed the equipment for removing carbon dioxide from power stations and industry fifty years ago we would now face surmountable problems. There would still be the need to sequester carbon dioxide from the 30 per cent of emissions that come from all forms of transport – aircraft, cars, buses, trains, ships and trucks – but that might be done through gradual replacement. Now, ironically, we have ample supplies of carbon energy still available in the vast underground deposits of crude oil and even larger deposits of coal and tar sands, but a growing realization that we dare not use them in the carefree way of the last century.

The world's annual production of carbon dioxide is 27,000 million tons. If this much were frozen into solid carbon dioxide at –80°C it would make a mountain one mile high and twelve miles in circumference. To sequester this much each year could not be achieved quickly – probably not sooner than twenty years from now. With all the will and enthusiasm it still takes about twenty to forty years for any new technology to become global. This was true of steam power, electricity, aviation, radio and television and computers.

It is important to keep in mind that the ability to use comparatively cheap fuel is not limited by the availability of crude oil. The oil industry may move from crude oil to gas and coal as the feedstock from which they produce their products: petrol, diesel and aviation fuel. These products will continue to be used but, increasingly, they will be made from gaseous and solid fossil fuels. Chemists can devise ways to make them and other less polluting fuels for transport, from any energy source, even nuclear, but such is the inertia of industrial civilization that we are likely to go on using fossil fuel for a decade at least.

The most gloomy thought is the likelihood that we are unable to stop emissions in time; think how difficult it could be for those large nations China, India and the United States to overcome the social inertia of their massive populations. Whatever happens, we have to give up fossil fuel as soon as possible, because even when we are past the threshold of irreversible climate change, the extent and rate of adverse change will still be affected by what we do. Our aim should now be to try for the least hot future world.

Natural Gas

Natural gas in many ways seems an almost ideal fossil fuel and it is used to produce electricity in gas turbine power plants that are compact, highly efficient and can be built in or near centres of population where they are a combined source of heat and power.

Governments and industries wishing to reduce emissions of carbon dioxide and so lessen their guilt in global warming have welcomed the chance to burn natural gas instead of coal or oil. The main constituent of natural gas is methane, the simplest of the hydrocarbons, with just one carbon atom and four hydrogen atoms in its molecule. For the same energy output as from coal or oil, methane combustion releases only half as much carbon dioxide. This implies that powering a nation entirely by gas reduces emissions of carbon dioxide by half. What a wonderful way of meeting the target set by international agreements like Kyoto.

Unfortunately, in practice some of the natural gas leaks into the air before it is burnt. According to the Society of Chemical Industry's report in 2004, this amounts to about 2 to 4 per cent of the gas used. Along the many thousands of miles of pipeline that convey gas from the production sites to the power stations and homes, leaks do happen in spite of great care. The greatest leakage is usually at the production sites, though some of it leaks in our homes where it is burnt. Every time the gas is lit, some escapes into the air and is not burnt and when a gas flame is turned off, the unburnt gas leaks from the pipe linking the tap or gas valve to the burner. There are millions of homes using gas for cooking and central heating, and the leaks, although small individually, add up to a significant part of the escape of methane to the air.

The problem with these escapes of methane is that this substance is twenty-four times more potent a greenhouse gas than carbon dioxide. Fortunately, it has a relatively short residence time in the air, and about 8 per cent of it oxidizes naturally each year. In twelve years, only 37 per cent of any methane escape remains, the rest oxidizing to carbon dioxide and water vapour. Carbon dioxide stays much longer in the air and has a complicated removal with an effective residence time of between fifty and a hundred years. About half of the carbon dioxide we have so far added to the air remains there.

But methane is still a cause for concern. If approximately 2 per cent of natural gas used each year leaks before burning, it causes over a period of twenty years a peak global warming equal to that of burning coal instead of natural gas; at a 2 per cent leak, the Kyoto advantage of gas is lost for much of the next two decades. If 4 per cent of it leaks, the greenhouse effect peaks at more than three times greater than that of burning coal. The claim that burning natural gas halves the emission of greenhouse gases for the same energy production as coal is therefore only true if there are no leaks anywhere, from the production source to the combustion chambers.

It is difficult to find estimates of the leakage of natural gas. A recent brief communication in *Nature* in April 2004 by J. Lelieveld and colleagues at the Max Planck Institute for Chemistry in Mainz, Germany, reports leakage from the Russian natural gas pipelines of 1.4 per cent and finds it comparable with the 1.5 per cent leakage reported for the United States. The German report does not provide an estimate of the leakage at the production sites or when the gas is burnt, and perhaps this is why it is lower than the 2 to 4 per cent leak rate given by the Society of Chemical Industry in 2004. This is a serious gap in our knowledge, and there should be a section of the IPCC specifically responsible for estimating methane leaks and considering how they might be prevented.

The problem of methane leaks is made worse by the chaotic nature of the political world, where gas production sites are often in unstable regions and proper supervision is near impossible. To a terrorist group a pipeline is an easy target, something utterly vulnerable, going thousands of miles across an open landscape. They know that a few pounds of Semtex exploded against the pipeline can cause huge damage to a national economy; when they realize the global threat of leaking gas they will have an even more powerful card to play in their blackmail of the world. In the present world it is conceivable that burning gas instead of coal could worsen not improve our chance of curbing global warming.

We have also to consider that soon the world will be shipping liquefied natural gas in giant tankers travelling from remote production sites to the hungry power stations in the USA, Japan, China and Europe. Methane becomes liquid at −160°C and can be shipped in vast insulated containers. Enough heat leaks in through the container walls to keep the liquid methane boiling, and some escapes; the longer the journey the greater the leakage. (Of course, tanker operators will use some of the escaping methane to fuel the tanker's engines.) And tankers do have accidents, and should this happen much of the liquid gas cargo would escape into the atmosphere.

HYDROGEN

Hydrogen, like electricity, has to be made; there are no hydrogen wells anywhere on Earth. Chemical engineers can design plants to make hydrogen from almost any of the other fuels – gas, oil or coal – or it can be made, using electricity, directly from water. Nuclear energy can be a source of hydrogen too, either through electricity made at a power station or directly using a high-temperature nuclear reactor. It is not difficult to make hydrogen, but I think it extremely unlikely that it will soon be flowing to homes and industry as a replacement for natural gas. It is unlikely that hydrogen will ever be distributed on any significant scale as a fuel for transport, and if it were feasible the infrastructure needed to make, transport and deliver hydrogen would take more time to build than we can spare.

We have a chance of evading some of the consequences of the second law if we burn fuel in a fuel cell. A fuel cell is no more than a battery having at one electrode the fuel and at the other electrode oxygen. In theory such a battery could convert most of the energy of the reaction into electricity – never 100 per cent, but certainly much better than achieved at even the best conventional power station now. So far, the best fuel to use is hydrogen, and the modules of the astronauts who went to the moon were all heated and supplied with electricity made by reacting hydrogen with oxygen in fuel cells. Fuel cells work, but they are at present expensive, and somewhat temperamental.

Hydrogen is much more difficult to handle than the other gaseous fuels, methane and propane. At present it can only be stored at very high pressures in strong metal or carbon fibre containers. It tends to make steel brittle, and because of its small molecular size it leaks easily through tiny holes that would be only a minor problem with a heavier gas such as propane. A hydrogen air mixture detonates when ignited, instead of burning fast but smoothly as does a methane air mixture. A hydrogen flame is invisible, so that the ignition of a small leak can cause dangerous overheating of piping or valves before it is noticed. All of these drawbacks can be overcome by good engineering, but they add to the time and cost of establishing a hydrogen economy.

There is a practical way to use hydrogen, and it has been described by the American engineer Geoffrey Ballard. He offers an intriguing hydrogen economy that does make sense. He foresees a national stockpile of hydrogen existing in the gas tanks of all the nations' cars. It would be used in the fuel cell that powers the car, but at the same time the hydrogen in the cars' tanks would be a national energy store.

Fuel cells are reversible; they can use hydrogen to make electricity efficiently or, equally, they can be a source of hydrogen when supplied with electricity. Ballard observes that the total fleet of cars in most nations has a generating capacity many times greater than the nation's power stations. All that is required is that each car, when it is not being driven, is plugged in to the national grid. This could be either at home or at car parks.

Cars and trucks would then be the fuel store and generators of the national electricity supply and be able to draw energy from it. The primary source of electricity would be power stations that did not emit greenhouse gases. The hydrogen would be part of millions of small storage batteries disbursed everywhere. This to me is an imaginative and attractive scheme, and I hope that the technology of fuel cells and hydrogen storage vessels develops until it can be done.

RENEWABLES

The phrases 'sustainable development' and 'renewable energy' have entered the language of politics and politicians use them to establish their concern for the environment and their green credentials. I doubt if Dr Gro Harlem Bruntland, who first introduced the concept of sustainable development, ever imagined that her good intentions would be so grievously misunderstood. I wonder if she feels as I did when, in Japan two years ago, I saw a car called 'Gaia'. It was not even a hybrid vehicle designed to be energy saving.

Before the twentieth century ended, we were unaware how serious a threat global heating was, and we believed that civilization could only flourish if there was unceasing economic growth. Some of us doubted this economic dogma and strived for ways to uncouple economic growth from the consumption of fossil fuels and raw materials. Few were better at this than Amory Lovins, and the members of his Rocky Mountain Institute, whose thoughts led to the invention and manufacture of much less polluting cars, the hybrids, part electric and part internal combustion, and to a cornucopia of ingenious yet

Torrents of water pouring from Greenland's melting glaciers. Sea levels are currently rising 6 centimetres a decade, partly as a result of the melting glaciers.

Exit Glacier, Harding Icefields, Alaska. The post shows where the glacier reached in 1978, since when it has retreated half a mile. Similar glacier shrinkage is occurring globally.

Dumai, Indonesia. A farmer tries to extinguish a peat bog fire. In 2002 these fires contributed 40 per cent of the world's total carbon emissions.

Dorset, England. Pre-agribusiness English countryside.

Amazon rainforest, Anapu Region, Northern Brazil. Widespread deforestation is leading to the loss of one of the world's great ecosystems. The Amazon shrinks by approximately 20,000 square kilometres (the size of Wales) every year.

How we are changing the face of the Earth to feed six billion people. It is as much a part of global heating as greenhouse gas emissions.

Algal life in the oceans – abundant coloured red and yellow, sparse coloured blue and purple and showing the tropical ocean deserts. Ocean algae are essential for climate regulation.

A composite photograph showing the earth at night as seen from space, illustrating the massive amounts of energy used by humans and the spread of urban areas.

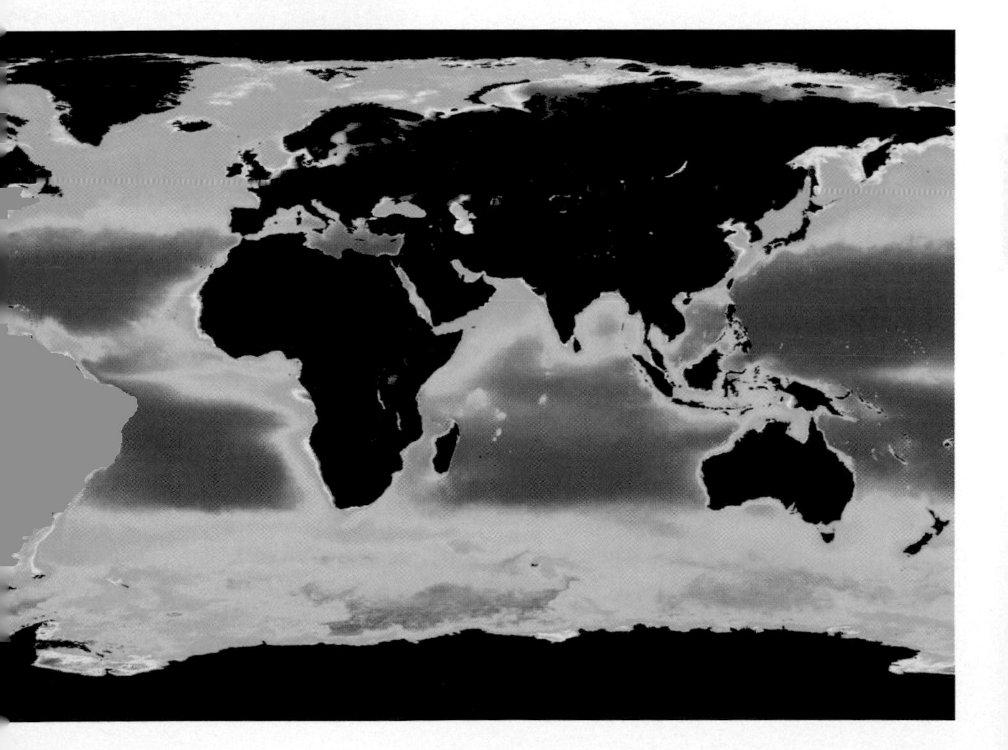

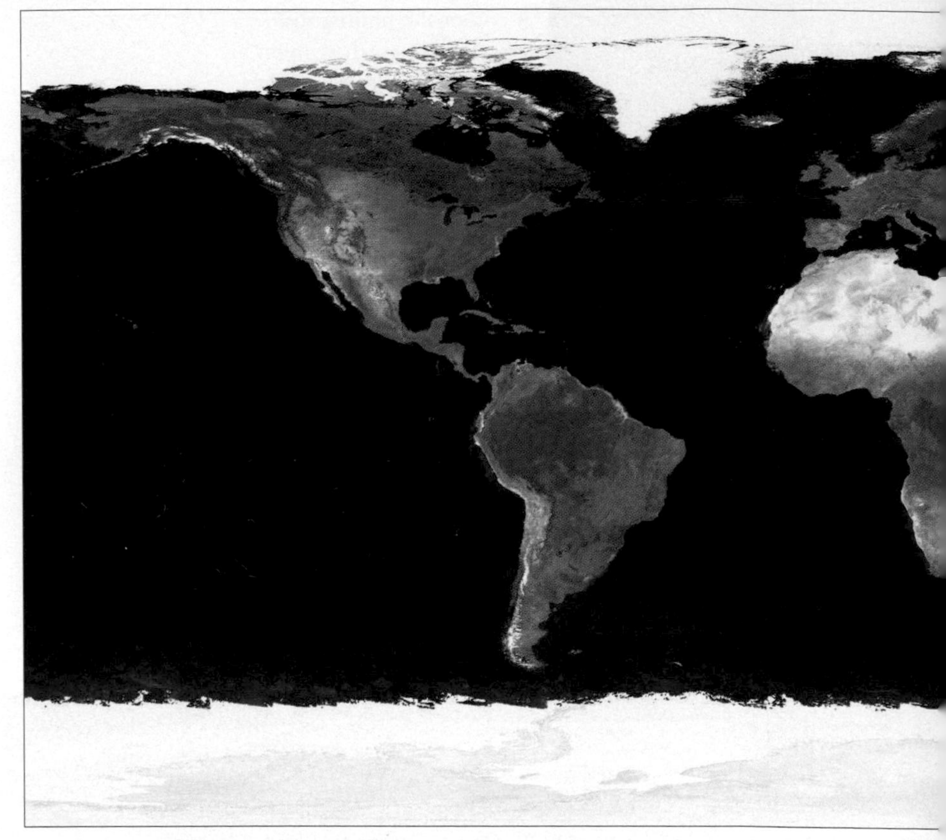

A composite photograph which shows the relatively small proportion of the earth's surface covered by vegetation.

Mars now – and what the earth will look like eventually.

Cornwall, England. Land devastated by tin and copper mining.

The acceptance of nuclear waste by the natural world – Par Pond, a waste repository at the Savannah River Nuclear Facility in the USA.

practical energy-saving devices. Their notion of economic growth without environmental cost was part of the inspiration for the concept of renewable energy.

As long ago as 1981, Stephen Schneider featured as a chapter epigraph in his book *The Primordial Bond* Paul and Anne Ehrlich's warning:

> The environmental system of the Earth would collapse if the attempt were made to supply all human beings alive today with a European style of living. To suggest that such an increase in living standards is possible for a world population twice the present size by the early part of the next century is preposterous.

Since the beginning of the nineteenth century we have taken more from the Earth than it could provide. Sustainable development and renewable energy might have worked in earlier times, but I think that to expect them and energy saving to sustain our numbers today is no more than a romantic dream.

Europe has seriously damaged its countryside and its competitiveness in the world by a byzantine mix of subsidies, credits and horse trading called the common agricultural policy (CAP). It now seems hell bent on an even madder common energy policy. What was left of the German landscape has been diminished by becoming the site for 17,000 huge wind turbines. The UK is fast following the German example, as Denmark has already done.

Since Europe is so deeply committed to wind energy and many of its nations are strongly opposed to nuclear energy, I am saying more about wind and nuclear fission in this chapter than about other energy sources. One of the best, although partisan, sources on the various forms of renewable energy is a book edited by Godfrey Boyle, *Renewable Energy*, which was a major component of the Open University's undergraduate course on the subject. Alternatively, a visit to the Centre for Alternative Technology (CAT) at Machynlleth in Wales gives an immediate acquaintance with renewable energy sources.

Wind power

The never-ceasing movement of the world's air and oceans is driven by the heat from the sun. It happens because nearly all fluids, and certainly air and water, change density with any change of temperature. As the land warms in the equatorial sunlight, the air in contact also warms and becomes less dense; it then rises like a hot air balloon. When the sea surface warms it becomes lighter and floats on top of the cool waters below; it does not mix with these cooler waters and consequently a stratified warm upper layer forms. From the warm ocean surface water evaporates continuously and mixes with the air, lowering its density and providing what is called insensible heat, so called because it is a measure of the heat content of the air, not its temperature. It takes an astonishing amount of heat to evaporate a gram of water – about 600 calories – and this heat is retrievable when the water condenses again. As a parcel of warm, wet air rises it cools and the water vapour condenses, releasing its latent heat and providing more heat energy for the air parcel to rise further. This is part of the force that energizes the tropical thunder storms.

The next consequence of all this rising warm damp air is movement away from the equator to the north and to the south. The vertical motion of the warmed air cannot easily penetrate into the stratosphere, a layer of the atmosphere that exists, in north temperate regions, above about ten kilometres from the surface (where most of us have been without knowing it, as passengers travelling in jet aircraft). The stratosphere is warmer than the air mass immediately below, and lies above it much as the warm water of the surface layer of the ocean lies above the cooler water beneath. The boundary between these two separated parts of the air is called the tropopause, and it forms an invisible barrier to upward air motion. It is higher, at about seventeen kilometres, in equatorial regions. The warm wet tropical air rises through the troposphere – the lower atmosphere where we and the clouds exist – and as it rises sheds its water as rain. When the dried air reaches the tropopause it turns north or south and moves as a pair of flat cylinders girdling the planet. When the northerly or southerly motion takes the air to about latitudes 30° north or south it begins to descend; the downward movement now heats the air by compression and makes the surface regions in the descending air the hottest and driest parts of the world: the deserts of Australia, Chile, Sahara, Texas and Mexico, and the Persian Gulf.

George Hadley first proposed this form of planetary air motion in a paper to the Royal Society in 1735 entitled 'Concerning the cause of the General Trade Winds', and the cellular zones of the Earth where it happens are now called Hadley cells. His name was rightly chosen for one of the foremost climate centres of the world, now part of the Meteorological Office at Exeter, in the UK, which was established by the then Prime Minister Margaret Thatcher in 1988. Global warming tends to make the Hadley cells grow larger and extend further north and south. The hot dry regions could then extend into the temperate zone. The return flows of dry air to the tropics are the north-east and south-east trade winds so welcomed by sailors. Because of the expansion of the Hadley cells as global heating intensifies, it would be unwise for Europeans to assume that the now prevailing westerly winds will continue to blow at the same latitude.

Air motion driven by heat is the source of the wind, but the motion of a fluid is rarely ever simple; the water in your kitchen sink never flows uniformly into the waste pipe when the plug is pulled. It often forms a vortex, spinning as it goes. Sometimes this vortex is powerful enough to have a central empty core, down which air is noisily dragged along with the water. So it is with the atmosphere: the large-scale heat-driven motions that spin as hurricanes or cyclones are the source of the wind. There is no simple rational answer to the question 'why does a vortex form?' All that we can say is: whenever there is a flow of energy through matter, interesting things like vortices, flames and life emerge. Erich Jantsch, in *The Self-organizing Universe* (1980), observed that we seem to live in a universe where orderly structures form whenever there is a flow of energy.

Humankind has used wind energy from its earliest times, mainly to move wooden sailing ships across the sea. The green movement has been the advocate for this constant clean source of energy, and in the long run it might have promise. But at present, wind energy as a whole system is in the early stages of development and is not much more efficient than were those early biplanes held together with wire that were the first form of air transport. We still have much to learn about using wind power, and most of all about storing the energy it produces when the wind blows. The wind is restless and blows only some of the time. Anticyclones with little or no wind bring the hot days of summer when air-conditioning may be needed, and they bring the cold frosty days of winter when energy is required to keep warm. But when we can store the wind energy, all could be well. There is no reason in principle why we can't; there is, for example, in Wales a high reservoir built in the 1950s into which water is pumped using electricity when it is in surplus, such as in the middle of the night. When electricity is needed such as at rush hour, water is drawn from this reservoir through water turbines and used to generate the needed extra supply of electricity. This is potentially a fine and reliable way of storing energy but it needs a suitable mountainous region close to a windy place. Other ways of storing energy, such as using compressed air, can be thought of and planned, but the loose talk of enthusiasts for wind energy that the energy could be stored as hydrogen and that this could then be used as fuel for cars, ignores the decades of engineering development needed to make this a practical option. At present none of them is immediately available on the scale that is needed.

There are many parts of the world, the Great Plains of the United States and Russia, for example, where wind farms could coexist with fields of agribusiness and be welcome; wind farms offshore sound good, too, as the wind is more powerful and reliable than on land and they could be out of sight. Unfortunately, the costs of maintenance are much higher than for land-based turbines. Each individual turbine would have to be serviced by small boats; unfavourable tides and rough seas would often delay or prevent them from docking at the turbine. The placement of efficient, that is, huge (100 metres or higher) wind turbines in the densely populated parts of Europe is proving highly unpopular. On aesthetic grounds these are not suitable places for harvesting wind energy on a large scale.

Aesthetics alone is an insufficient reason for rejecting what might be a clean and valued energy resource, and if wind power was truly capable of providing a serious proportion of our energy needs in Western Europe, most of us would grit our teeth and accept it, even though it is to many an unpleasant and intrusive power system. It is sometimes claimed by wind enthusiasts that all our electricity could come from wind; I doubt if many of them have calculated the number of 100 metre, one megawatt turbines needed. To supply the UK's present electricity needs would require 276,000 wind generators, about three per square mile, if national parks, urban, suburban and industrial areas are excluded; also needed would be an efficient way of storing the electricity they produced. But in no way is it efficient and economic; the intermittency of the wind means that, at best, energy is available from wind turbines only 25 per cent of the time. During the remaining 75 per cent, electricity has to be made in standby fossil fuel power stations; worse still, the power stations have to be kept idling when wind energy is available, an inefficient way for them to operate. The most recent report from Germany put wind energy as available only 16 per cent of the time, and in Denmark, which has pioneered their development, Niels Gram of the Danish Federation of Industries said, 'In green terms windmills are a mistake and economically make no sense . . . Many of us thought wind was the 100 per cent solution for the future but we were wrong. In fact, taking all energy needs into account it is only a 3 per cent solution.'

According to the Royal Society of Engineers 2004 report, onshore European wind energy is two and a half times, and offshore wind energy over three times, more expensive per kilowatt hour than gas or nuclear energy. No sensible community would ever support so outrageously expensive and unreliable an energy source were it not that the true costs have been hidden from the public by subsidies and the distortion of market forces through legislation. Enthusiasm for renewable energy coupled with a politics in which each nation tries to gain brownie points for its diligence in meeting the Kyoto limits is an unhappy mixture. It will fail and bring discredit both to the greens and to the politicians foolish enough to adopt renewables as a major source of energy before they have been properly developed.

Wind energy, through crude and unsustainable industrial development, is already devastating some unusually beautiful countryside. That countryside, although already damaged by agribusiness, still has a few areas that are an example of how to live in a decent and seemly way with the natural world. I think that the responsibility for the wrong advice given to the government came from well-meaning city dwellers with a romantic, impractical dream of clean renewable energy coupled with a misplaced fear of nuclear energy but no real empathy with Gaia or the natural world. We might have been wiser to seek to use the energy of the ocean in the form of waves and tides.

Wave and tidal energy

Tidal energy makes use of the stored gravitational energy of the Earth, Moon and Sun system. Science fiction tells of future civilizations who draw almost all their energy needs from this clean renewable source; the consequence is the eventual decay of the Moon's orbit until, close to the Earth and filling the sky, the moon is torn apart by the uneven pull of gravity. This need not deter us from starting now to build modest tidal power schemes. I am indebted to Jonathon Porritt, the authoritative leader of environmental opinion, for details of a tidal energy scheme for the Severn Estuary; they were strongly supported in a statement by Professor Ian Fells at a meeting at Dartington Hall in Devon in June 2004, who said that a Severn Barrage was estimated to cost £13 billion, but since it could provide 6 per cent of the UK's energy needs, it was an attractive business proposition. At La Rance near Cherbourg in France a similar but smaller tidal energy scheme has operated for many years now and provides energy to supplement the mainly nuclear French electricity supply.

There are several experimental schemes now running around the coasts of the UK that aim to draw energy from the movements of the sea. Some use the motion of the waves, others the tides and still others the currents that flow in the sea as a consequence of the tides. An excellent review of tidal energy is in *Chemistry and Engineering News* for October 2004. While such schemes seem well worth while as experiments and to gain hands-on experience, we should not expect even the most promising of them to deliver a substantial part of our energy needs before at least twenty, and more probably forty, years have passed.

It is rarely appreciated that almost every engineering development, whether steam power, electricity supply, radio, television, telephones or passenger aircraft, took about forty years to pass from open enthusiasm to widespread application in the first world. I see no signs that this gestation period can be lessened, except perhaps when the imperatives of war cause a whole nation to act in unison.

Hydro-electricity

Water mills are probably our oldest renewable energy source, and hydro-electricity is now a matured and significant source of energy. Some nations, for example Canada, Norway and Sweden, satisfy as much as half their energy needs by water power. China has recently built the largest hydro-electric plant of all: the Yangtze dam supplies sixteen gigawatts of electricity. Although not free of dangers and environmental disturbance, hydro-electricity is far less damaging than fossil fuel combustion. Unfortunately, there are too many of us and too few rivers in Britain and in many other parts of the world, for this benign energy source to satisfy more than a small fraction of our total consumption.

Bio fuels

Used sensibly and on a modest scale, burning wood or agricultural waste for heat or energy is no threat to Gaia, but we have to remind ourselves that bio fuel, when harvested in a large-scale operation, is a menace. It is only renewable if it has no effect on the natural cycle of carbon. Bio fuels are especially dangerous because it is too easy to grow them as a replacement for fossil fuel; they will then demand an area of land or ocean far larger than Gaia can afford. If it is perverse and dangerous to gain energy by burning fossil carbon in fossil oxygen, it is equally so to imagine that comparable quantities of energy are freely and safely available from this much-applauded renewable energy source. We have to discard the old-fashioned teaching of both science and religion and begin to look on the forested land surface of the Earth as something that evolved to serve the metabolism of the Earth; it is irreplaceable. We have already taken more than half of the productive land to grow food for ourselves. How can we expect Gaia to manage the Earth if we try to take the rest of the land for fuel production?

Solar energy

No wonder many of the ancients worshipped the sun; it is the ultimate source of everything needed by life on Earth. Not only does it keep us warm by the unceasing supply of 1.35 kilowatts of energy for every square metre of surface on which it shines; it also supplies the light that enables the photosynthetic production of living organisms, and it ultimately feeds us and gave us our fossil fuels. Most of all, the sun empowers Gaia to self-regulate our planet.

Why on earth, you may ask, can't we use solar energy directly? It must amount to far more than even our present needs.

It is certainly possible to convert sunlight directly into electricity in many different ways. We can concentrate sunlight by focusing it with large lenses or mirrors and use the heat to power a steam engine connected to a generator. We can take electricity directly from solar cells; these are usually made of silicon, the element that made possible the many different electronic devices we use every day. These solar cells cause the high-energy particles, photons, of sunlight to detach electrons from the silicon crystals, and the flow of these electrons is the electric current that the solar cell provides. Solar cells are invaluable and provide energy for the numerous man-made satellites that circle the Earth and serve to provide near-instantaneous information transmission, television on a global scale and the monitoring of the air, sea and land. They are similarly used at remote places, on mountains and in deserts, far from the copper wire of the electricity supply.

But solar cells are not yet suitable for supplying electricity directly to homes or workplaces, mostly because, despite over thirty years of development, they are quite expensive to make. At the Centre for Alternative Technology in Wales there is an experimental house with a roof made almost entirely of silicon photocells. In summer it provides about three kilowatts of electricity, but the cost of the installation was comparable with that of the house itself, and the expected life of the cells is about ten years. Sunlight, like wind, is intermittent and would, without efficient storage, be an inconvenient energy source at these latitudes. A great scientific effort is under way to produce affordable solar cells from material such as plastics that could be manufactured in bulk. So far as I know, the production of a solar cell which is inexpensive, long-lived and which efficiently converts sunlight into electricity has not been achieved, and there are no large-scale economic and attractive photoelectric sources that could be used for medium- or large-scale energy supply. This is especially true for northern temperate regions, where the sun in winter is low in the sky and the weather often cloudy.

If an economic, 25 per cent efficient converter of sunlight into electricity could be made available as roofing material it would provide a fine and sensible energy supplement. But as with wind, the intermittency of solar supply would necessitate efficient energy storage, and so far this too is unavailable. I find it hard to believe that large-scale solar energy plants in desert regions, where the intensity and constancy of sunlight could be relied on, would compare in cost and reliability with fission or fusion energy, especially when the cost of transmitting the energy was taken into account.

NUCLEAR ENERGY

There are at present two quite different sources of nuclear energy. The first, nuclear *fission*, uses the energy released when the large atoms of elements such as thorium, uranium and plutonium split apart. Fission energizes the present-day nuclear power stations of the world. It also powers nuclear submarines and provides the explosive force of nuclear weapons. The other source of nuclear energy is the *fusion* of the nuclei of light elements, such as hydrogen and its isotopes. This energy powers the sun and most other stars; it is not yet providing electricity for public use but does provide part of the explosive energy of the 'hydrogen bomb'. Provided that engineering problems do not prevent the building of practical and efficient fusion power stations, I think that these will be the future source of electricity.

Fusion Energy

When hydrogen gas burns, the flame is hot and it provides enough energy for it to be considered as a possible fuel for cars and other vehicles. The energy of hydrogen combustion comes from the movement of an electron in orbit around the hydrogen atom to a vacancy in the shell of electrons surrounding an oxygen atom. This motion of the electron can be regarded as a tiny electric current, and the potential that drives it is 0.82 volts; this current flow when the countless trillion of hydrogen atoms burn as a flame is what keeps it hot. To start hydrogen and oxygen burning, the mixture has to be heated to over 500°C, at which temperature the gas molecules are moving fast enough for a sufficient proportion of collisions to be consummated and allow enough heat for the reaction to become self-sustaining. If we could heat hydrogen atoms to well over 150 million degrees, their speed would be so fast and collisions between them so hard that a few of the collisions would cause the atoms to fuse together to form the heavier atom helium; the act of fusion releases a prodigious quantity of energy, as much as would come from the impact of an electron accelerated by a potential of 21 million volts. This means that the nuclear fusion of hydrogen yields millions of times more energy than its mere combustion, but to start the powerful reaction requires some means of heating the hydrogen to 150 million degrees.

Along with many other scientists throughout the world, I knew that nuclear fusion energy, the nuclear combustion of hydrogen, was the ultimate clean and everlasting energy source, mainly because we knew that this was what empowered the sun and other stars. Most of us still thought that we were a long way from realizing fusion in practice. It just seemed impossible that the conditions inside the core of the sun, with temperatures over 100 million degrees, could be arranged here on Earth on a practical scale as part of a power station.

But in February 2005 the director of the Culham Science Centre, Professor Sir Christopher Llewellyn Smith, invited Sandy and me to visit and view their Tokomak reactor, and to learn about their recent experiences using it and the prospects for fusion energy. We were amazed and delighted to discover that their fusion reactor had proved itself by sustaining for two seconds a nuclear flame that burnt deuterium and tritium, isotopes of hydrogen, and generated sixteen megawatts of energy. Admittedly it was only 64 per cent of the energy needed to ignite the flame, but it proved that the physics and the engineering were sound and worked as expected. The Culham reactor was the prototype from which a pilot power station could be designed and then the first working fusion power station.

As a scientist I was intrigued by the thought that there in front of me was the large toroidal flask within which temperatures far above those of the hottest part of the sun's core had been sustained for a couple of seconds. The temperature of the burning mix of hydrogen isotopes was 150 million degrees, which compares with the 100 million degrees at the centre of the sun. The sun, of course, can afford to burn at a much more leisurely pace.

The deuterium fuel used for fusion energy is unlimited in availability. It constitutes 0.016 per cent of water and it is easy to extract. The second fuel, the radioactive hydrogen isotope tritium, has to be manufactured. In the strange world of nuclear energy, tritium is produced by the fusion reactor as it runs. As the two hydrogen isotopes fuse they generate energy in the form of two energetic particles, one is a helium atom with three million electron volts and the other a neutron with fourteen million volts of energy. The kinetic energy of the helium atom provides the heat that keeps the plasma flame hot, and the neutrons give up their huge kinetic energy in the walls of the reactor, where it degrades to heat. In a future power reactor, the heat from the neutron flux could supply the thermal energy to gas or steam turbines that would then make electricity. The neutron flux could also provide a constant source of new tritium fuel by its reaction with a lithium isotope incorporated in the reactor walls.

The nuclear waste of a fusion reactor is the harmless non-radioactive gas helium, and there are no long-term radioactive wastes. The metal parts of the reactor become mildly radioactive as a consequence of the neutron flux, but this is a minor disposal problem.

So why are we not now supplied with safe fusion energy? It is because the magnificent international developments at Culham have gone about as fast as can reasonably be expected; the power generated in fusion reactors has grown over the last twenty years at a pace greater than the rate at which computers have increased their speed and capacity. It is now almost as large as that needed to start fusion, implying that a working fusion reactor is now within reach. This is an outstanding outcome, and we left Culham with the thought that the next large thermonuclear reactor, due to be built in France, will be producing power for a national grid. It will be the prototype for a growing number of safe, reliable energy producers.

If Kyoto had been influenced more by the pragmatism of scientists and engineers and less by romantic idealism, we might soon have harvested fusion energy. As it is, even given good will, it may take twenty more years before it begins to heat our electric kettles or run our word processors.

Fission Energy

As with the other energy sources, I am not providing here information on the construction of the different designs of nuclear power plants or on the scientific concepts that form their basis; what I am discussing are the merits of nuclear energy as a Gaia friendly energy source and its safety. An excellent recent book by W. J. Nuttall, *The Nuclear Renaissance* (2005), is a fine starting place for anyone who wants to know more about the history, the hardware, and the politics of nuclear fission and fusion. An earlier review by Walt Patterson, *Transforming Electricity* (1999), is another starting place.

Many of my friends among environmentalists are surprised at the strength of my support for nuclear energy and seem to think that I have recently changed my mind. This is quite untrue as a glance at Chapter Two of my first book, *Gaia* (1979), and Chapter Seven of my second book, *Ages of Gaia* (1988), will show.

A television interviewer once asked me, 'But what about nuclear waste? Will it not poison the whole biosphere and persist for millions of years?' I knew this to be a nightmare fantasy wholly without substance in the real world. I also knew that the natural world would welcome nuclear waste as the perfect guardian against greedy developers, and whatever slight harm it might represent was a small price to pay. One of the striking things about places heavily contaminated by radioactive nuclides is the richness of their wildlife. This is true of the land around Chernobyl, the bomb test sites of the Pacific, and areas near the United States' Savannah River nuclear weapons plant of the Second World War. Wild plants and animals do not perceive radiation as dangerous, and any slight reduction it may cause in their lifespans is far less a hazard than is the presence of people and their pets. It is easy to forget that now we are so numerous, almost anything extra we do in the way of farming, forestry and home building is harmful to wildlife and to Gaia. The preference of wildlife for nuclear waste sites suggests that the best sites for its disposal are the tropical forests and other habitats in need of a reliable guardian against their destruction by hungry farmers and developers.

An outstanding advantage of nuclear over fossil fuel energy is how easy it is to deal with the waste it produces. Burning fossil fuels produces 27,000 million tons of carbon dioxide yearly, enough, as I mentioned earlier, to make, if solidified, a mountain nearly one mile high and with a base twelve miles in circumference. The same quantity of energy produced from nuclear fission reactions would generate two million times less waste, and it would occupy a sixteen-metre cube. The carbon dioxide waste is invisible but so deadly that if its emissions go unchecked it will kill nearly everyone. The nuclear waste buried in pits at the production sites is no threat to Gaia and dangerous only to those foolish enough to expose themselves to its radiation.

There is much loose talk of burying the carbon dioxide waste, but there seems to be little recognition of the sheer difficulty of the task. How is it to be collected from the myriad sources around the world? Where can we put these mountains that we make each year? I find it sad, but all too human, that there are vast bureaucracies concerned about nuclear waste, huge organizations devoted to decommissioning nuclear power stations, but nothing comparable to deal with that truly malign waste, carbon dioxide.

But it is not enough to use this as an argument favouring a wider use of nuclear energy, because the public belief in the harmfulness of nuclear power is too strong to break by direct argument. Instead, I have offered in public to accept all of the high-level waste produced in a year from a nuclear power station for deposit on my small plot of land; it would occupy a space about a cubic metre in size and fit safely in a concrete pit, and I would use the heat from its decaying radioactive elements to heat my home. It would be a waste not to use it. More important, it would be no danger to me, my family or the wildlife.

In the endless debate about nuclear energy, it is often assumed that an anti-nuclear David is bravely confronting a Goliath nuclear industry. What a false image this is. The green lobbies are large, whereas the nuclear industry, by comparison, is tiny compared with oil and coal companies, which can be very large indeed. A moment's thought on the power densities of carbon fuels compared with nuclear fuels explains why the nuclear industry is small. To produce the same amount of electricity requires a million times more oil or gas than it does uranium. As a consequence the nuclear industry can hardly afford pro-nuclear demonstrations and advertisements, and you rarely ever hear the counter-arguments.

Another factor that has sustained the false image of nuclear dangers is the reluctance of scientists to speak in public. A good scientist knows that nothing is certain; everything is a matter of probability; whereas an anti-nuclear activist will never hesitate to exaggerate and speculate. It needs little imagination to see how feeble a good and honest scientist can be made to appear in the adversarial atmosphere of a court room, or on television. Especially if, as so often in a media debate, the chairman is out for an entertaining fight, not mere dull information.

Even more than this, scientists today are hampered by their low social and economic status. Long gone is the respect and independence given to Lavoisier, Darwin, Faraday, Maxwell, Perkin, Curie and Einstein. Hardly any laboratory scientist anywhere is as free as a good writer can be. Indeed I suspect that the only scientists we know well are those who can write entertaining books; the real contributors to knowledge are mostly unknown. Younger scientists cannot freely express their opinions without risking their ability to apply for grants or publish papers. Much worse than this, few of them can now follow that strange and serendipitous path that leads to deep discovery. They are not constrained by political or theological tyrannies, but by the ever-clinging hands of the jobsworths that form the vast tribe of the qualified but hampering middle management and the safety officials that surround them.

So why are so many against nuclear energy? How did these false fears arise? I think they go back to the Second World War when President Truman – the man who said of difficult decisions, 'the buck stops here' – had the awesome task of deciding between dropping the newly minted nuclear bomb on a Japanese city or merely demonstrating its fearful power to the Japanese military.

That it was used to destroy Hiroshima and Nagasaki gave birth to a wholly new perception of things nuclear. We could no longer see it as a wondrous gift of pollution-free energy; our minds were clouded by fear of nuclear war, a fear that has persisted. We might have seen more clearly the benefit of nuclear energy had not America tried to keep what it saw as its own secret, and if the polarization of politics had not crystallized into a Cold War between capitalism, represented by the United States, and communism, represented by Soviet Russia. It was not long before the Soviet Union had its own bombs as well. Soon there was a nuclear arms race, and weapons grew ever stronger and most of us feared a war that would destroy not just the combatants but civilization as well. It was in this pathological world state, wracked by feverish confrontations like the Cuba crisis, that anti-nuclear protest began.

In *Nuclear Renaissance*, Nuttall offers the best account I have yet read of the growth of anti-nuclear feeling in western democracies.

The real opposition to nuclear power within the public grew in the 1970s and the 1980s. It may be argued that this has been a consequence of the rise of single-issue pressure groups and youth culture. That is, as the anti-Vietnam War demonstrations of the late 1960s grew out of earlier Civil Rights demonstrations, so the anti-nuclear demonstrations of the late 1970s arose directly from the Vietnam War protests, once that conflict had come to an end. This, however, is a rather Americanized perspective on what has been an erosion of enthusiasm for nuclear power. In Britain the defining socio-political events of relevance are those associated with the Campaign for Nuclear Disarmament (CND) in the late 1960s and resurgently in the early 1980s. Not only was CND passionate and anti-American, but it was also fun and it was cool. This fusion of popular culture with the British anti-nuclear movement of the 1960s is vividly captured by the present writer's uncle Jeff Nuttall in his visceral autobiography *Bomb Culture*, in which he describes one CND Aldermaston March as a Carnival of Optimism: 'Protest was associated with festivity.' This important aspect of matters nuclear has only slightly attenuated with passing decades. Those advocating nuclear renaissance ignore such aspects of nuclear power at their peril.

I agree with Nuttall, and it is easy to see why many greens are so anti-nuclear; they often are the children of a union between environmentalism and the CND.

Before the Cold War intensified in the late 1950s there was widespread hope that nuclear energy was good and could play its part in reconstructing a decent civilization. In the United Kingdom, one of several European nations where the science of nuclear fission was born in the 1930s, our Queen opened in 1956 the world's first nuclear power station at Calder Hall. It was an event welcomed almost everywhere. The euphoria did not last; gradually as the Cold War intensified and the two superpowers tested larger and ever larger weapons, the all-pervasive fear of all things nuclear became widespread. This period of madness culminated in 1962 in the test explosions of hydrogen bombs equal in power to 20,000 of the bombs dropped on Hiroshima. The superpowers were rattling the Earth to show how strong they were, strong enough for mutually assured destruction. Mad it may have been, but it showed that each superpower now possessed the capacity to destroy civilization.

There were several interesting consequences of these vast explosions. They released into the global atmosphere radioactivity as great as that from two Chernobyl disasters every week for a whole year. The stratospheric winds carried the radioactive debris around the world and we all breathed in, or swallowed, such fission products as caesium 137 and strontium 90 and unexploded plutonium. Soon it was possible to demonstrate the presence of the strontium isotope in the bones of anyone in the world. Whatever harm to humans was done by these tests and their fallout, there is no evidence or theoretical conclusions to suggest that it held back the progressive increase in our lifespans; we now live longer than ever before – indeed European governments are now worried about how to pay the pensions of their ancient citizenry. It may comfort us to know that these tests, which produced as much fallout as a medium-scale nuclear war, posed no great threat to the Earth or to the health and well-being of its inhabitants.

An unscheduled benefit of the tests was the provision to Earth scientists of a suite of radioactive elements that could be used as tracers to follow the great natural cycles of the Earth system. From these tests we have gained a much deeper understanding of Gaia. Figure 6 illustrates the large and planet-wide radioactive contamination, almost all of it from bomb tests, during the last third of the century.

The ease with which even the tiniest traces of radioactivity can be detected and measured gave numbers that anti-nuclear activists could use to show that the 'poison' from these tests was everywhere in the world. They ignored the famous dictum of Paracelsus, 'The poison is the dose', and the fact that we ourselves are naturally much more radioactive than the fallout we had imbibed. The numbers were facts and the media did not hesitate to use them in their scare stories; perhaps they were justified because they made us reconsider the tests and conclude that they were a step too far. In 1992 the test ban moratorium ended all nuclear testing.

Among those appalled by the demonstrations by these two superpowers that they could destroy each other and perhaps civilization as

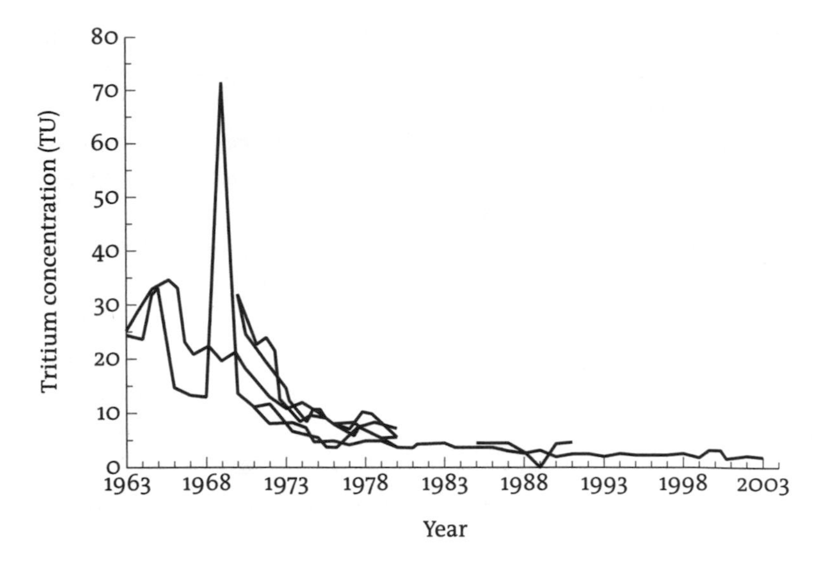

Figure 6. *Atmospheric radioactivity, represented by tritium, since 1963 when there were several large nuclear tests.*

well was the novelist Neville Shute. His 1961 book *On the Beach* was almost thermonuclear in its power, picturing the Earth totally destroyed by nuclear radiation. It was a wonderfully told story, wildly untrue, but it convinced many that all things nuclear are deadly. Among them was the Australian paediatrician Helen Caldicott, who became the most vocal and effective advocate of the worldwide anti-nuclear movement. Her advocacy led to the award of the Nobel Peace Prize, which gave enormous authority to her view of nuclear energy. In *Nuclear Madness* (1994) she wrote:

> As a physician, I contend that nuclear technology threatens life on our planet with extinction if the present trends continue; the food we eat, the water we drink will soon be contaminated with enough radioactive pollutants to pose a potential health hazard far greater than any plague humanity has ever experienced.

In the cause of deterring nuclear warfare between the superpowers, this cosmic-scale exaggeration by Helen Caldicott is excusable. But that was a twentieth-century problem, what we now face is much more deadly: the return to a new hot age. Ironically, if this happens, anti-nuclear advocacy will have hastened it.

For decades the fear of nuclear war was intense, and many novels and short stories were written, vying with each other to portray the nuclear nightmare. Hollywood joined in with its usual nonsensical hype, typified by the film *The China Syndrome*. In this, a badly constructed reactor disastrously goes wrong and a character in the film imagines its fissioning core melting its way through to the centre of the Earth, then continuing on miraculously until it emerges in China. Even as metaphor, this was a wholly absurd image, but it did its job of titillating public panic and fear and set the scene for endless misinformation and lies. These were enhanced a few weeks later by an expensive accident at the Three Mile Island nuclear power station in Pennsylvania. In spite of what you may have heard, the containment vessel of the reactor did its job and no one, either inside or outside the plant, was harmed.

Like a bright shining light, which broke through the fog of bad fiction and worse ideas, the wonderfully wicked comedy *Dr Strangelove, or, How I Learned to Stop Worrying and Love the Bomb* did something to restore our sense of humour and balance. True, it was singularly unfair to Edward Teller, the real Dr Strangelove, father of the hydrogen bomb. His autobiography reveals a good and peaceful man strangely equivalent to his Soviet counterpart, Andrei Sakharov. Few remember that Teller tried to persuade his government not to drop the first atom bomb on a Japanese city.

The nuclear bombs dropped on Nagasaki and Hiroshima were puny compared with the contemporary nuclear explosive devices borne by long-range missiles; each of these carries a clutch of separately targeted bomblets, each representing one megaton of explosive power, or about sixty-six Hiroshima bombs. A single one of these bomblets is powerful enough to ruin a major city. I can only imagine the consequences of their use in anger, but a visit to Hiroshima and its museum provides a glimpse of what could happen. I shall never forget my revulsion on seeing the way the light from that mere fifteen kiloton flash so illuminated and cauterized the town below that shadow images of people standing or sitting were melted on the surface of the stone walls behind them. We are right to be fearful of nuclear weapons. Perhaps their only virtue was that they frightened sufficiently the leaders of the superpowers, ensuring that the Cold War stayed cold for all its duration.

It is natural to fear the cancer that was the long-term consequence for some of those who escaped immediate death or wounding from the heat and the blast. So far, sixty years after the nuclear bombs in Japan, 7 per cent more of the survivors of the Hiroshima bomb have developed cancers than a comparable number of Japanese in parts unaffected by the bomb. In the affluent countries of the first world, the principal natural causes of death are heart disease, strokes and cancer; all three probably the result of growing old in an oxygen-rich atmosphere. The near 30 per cent who die naturally of cancer keep us all reminded of the grim prospect of malignancy. It is important to keep in mind that any increase in cancer attributable to all nuclear activity since the Second World War is still too small to be detected among the fluctuations in the natural death rate worldwide from cancer.

The twin fears of cancer and nuclear war are now ubiquitous in the developed world. In the underdeveloped world, not surprisingly, where death is more frequent and comes from overwork, malnutrition or disease, fear of radiation is much less. There is no time to think or fear cancer when the lifespan is only forty years, and when few experience the lingering misery of friends or relatives stricken with the disease, especially now that HIV makes personal tragedy commonplace.

Many Eastern European nations, once part of the Soviet empire, cling to their nuclear power stations even though they live in the shadow of Chernobyl; they see the benefits of nuclear energy far outweighing their alleged dangers. Fear is prevalent only in the pampered and cosseted developed world where there is a chance to live to ninety years or more. In that world money and research time are ceaselessly given in an effort to cure or prevent cancer and further to extend the lifespan. As someone now eighty-six, I am not much moved by this striving to live so long. I admit that if I could remain healthy and able to think at least as well as now, it would be good to sail on to one hundred years or more. The good life is surely not measured by its length in years, but by the intensity of the joy and good consequences of existence.

The irony of it all is that we in the developed world are the prime polluters, the most destructive of people on the planet, yet although we have the money and the means to prevent the Earth crossing the deadly threshold that will make global change irreversible, we are hampered by fear.

Chernobyl and the Safety of Nuclear Reactors

Franklin Roosevelt famously said on taking office in 1933, 'We have nothing to fear but fear itself.' Most of us have unreasoning fears which creep unwanted into the mind, and bring a shudder; mine are about overwhelming torrents of muddy water, at seeing and hearing a towering wall of water bearing down on me; something moving so fast that there is no chance of escape. I tell myself it is a foolish fear; we live high enough and far enough from the ocean that no conceivable tsunami would ever reach my home, and there are no great dams, filled with miles of water, upstream on our river. But still this nightmare scene steals into my dreams. I can well understand why many have similar fears of a nuclear catastrophe, fears that sensible explanation is never able to calm.

We need emission-free energy sources immediately, and there is no serious contender to nuclear fission. So how can we overcome our fear of nuclear energy? Remembering my own inconsolable fear of overwhelming torrents, it might be useful to compare the dangers facing two families – one living 100 miles downstream of the huge Yangtze Dam in China, a fine example of a powerful and effective source of renewable energy, and another living 100 miles downwind of the nuclear power station at Chernobyl, the worst example of the wrong kind of nuclear technology.

If the dam burst, perhaps as many as a million people would be killed in the wave of water roaring down the course of the Yangtze river. When the Chernobyl nuclear power station suffered a steam explosion and subsequently caught fire, releasing a large proportion of its radioactivity into an easterly air stream, the products were carried by the wind across much of the Ukraine and Europe. Many think that tens of thousands if not millions died as a result of the Chernobyl accident. As we will soon see, it was no more than 75.

I have never seen a dam burst or experienced in real life the terror that it would bring, but I have been in a cloud of radioactive nuclides escaping from a fire at a nuclear reactor. It happened in 1956, when the military reactor at Windscale in Cumbria caught fire and released a significant part of its accumulated activity into a northerly air stream, blowing down across England. At the time I was working as a scientist at the National Institute for Medical Research in north London. I was trying to discover, using the radioactive isotope iodine 131, more about the nature of the membrane of the human red blood cell. When I went to take my measurement I was annoyed to find that my primitive home-made Geiger counter was registering background beta radiation at a rate much higher than I expected of my samples, so measurement would be inaccurate if not impossible. My first thought was that my temperamental electronics were misbehaving, and I was about to start checking them when a colleague, Dr Tata, entered my lab and asked if I was having trouble measuring I_{131}. He and another scientist in the institute had found the background counts far above their usual level. Iodine is a volatile element, and we wondered if one of the three of us had accidentally spilt some radioactive iodine or flushed it unwisely into the laboratory sink. A few checks showed that I_{131} was everywhere throughout the building. We were all of us somewhat chastened and felt an unattributable sense of guilt. It was not until nearly twenty years later, on a visit to the Atomic Energy Authority's institute at Harwell, near Oxford, that I heard about the Windscale fire and the cloud of radioactive debris that contaminated most of England. In 1956, the year of the fire, the Government was able to keep the bad news bottled tight. They had the excuse that the reactor in question was part of the nuclear weapons programme and therefore steeped in official secrecy. The fledgling green lobbies and the media missed the chance to scare us all, perhaps even to death.

So far as I am aware, no one has reported any deaths or morbidity that could have come from the exposure of many millions of people to the release of 740 trillion becquerels of I_{131}. In the UK the National Health Service was a good record keeper, and any significant rise in the incidence of cancers would have been noticed. It was a real danger only to those at the scene itself, the firemen and the workers at the plant.

But that must be wrong, you say. Respectable media, for example the *Times* newspaper and the BBC, have more than once stated that 30,000 and more people have died in Europe and Russia as a result of exposure to radiation from the Chernobyl accident. I prefer to believe the physicians and radiobiologists of the UN agency the World Health Organisation (WHO). They examined the health of those in the area polluted by the plume from Chernobyl fourteen and nineteen years after the accident, and they were able to find evidence of only forty-five and seventy-five people, respectively, who had died. These were workers, firemen and others who bravely and successfully fought the fire in the burning reactor and carried out the cleanup afterwards.

So where do these false claims of a huge death toll from Chernobyl come from? They arise mostly from a perverse misinterpretation of the facts of radiobiology.

Careful and difficult observation and data gathering by epidemiologists has established a direct linear link between the dose of radiation received and death from cancer. Their data comes from the experiences of Japanese people exposed to the radiation from the atom bomb dropped on Hiroshima, from the use of radiation in medicine for both treatment and diagnosis, and from the life histories of radiologists and workers exposed to radiation during their working lives. The United Nations Scientific Committee on the Effects of Atomic Radiation (UNSCEAR) issued a report in 2000; this summarises the evidence and concludes that the hypothesis of a direct and linear response between radiation and harm done best fits the data. From the conclusions we could reasonably expect that the consequences of exposing the entire population of Europe to ten millisieverts of radiation, about as much as would come from 100 chest X-rays, would be 400,000 deaths.

Put like this it seems a terrible risk, but it is an amazingly naive way of presenting the facts. What matters is not whether we die but *when* we die. If the 400,000 were to die the week after the irradiation it would indeed be terrible, but what if instead they lived out their normal lifespans but died a week earlier than expected? The facts of radiation biology are that ten millisieverts of radiation reduces human lifespan by about four days, a much less emotive conclusion. Using the same calculations, the exposure of all those living in Northern Europe to Chernobyl's radiation on average reduces their lifespan by one to three hours. For comparison, a life-long smoker will lose seven years of life.

No wonder the media and the anti-nuclear activists prefer to talk of the risk of cancer death. It makes a better story than the loss of a few hours of life expectation. If a lie is defined as a statement that purposefully intends to deceive, the persistent repetition of the huge Chernobyl death toll is a powerful lie.

Chernobyl may well have cost some of those living in the Ukraine and Byelorussia several weeks of their life expectancy. How different it would have been had they lived on the flood plain of a river with a huge dam upstream that burst. Then they would have lost their whole life expectancy; this form of renewable energy can be much more deadly than nuclear.

A more solid and useful estimate of the comparative safety of the different energy sources comes from the Paul Scherrer Institute in Switzerland, in their 2001 report. They examined all of the large-scale energy sources of the world to compare their safety records. They expressed the danger of each in terms of the number of deaths from 1970 to 1992 per terrawatt year (twy) of energy made (a terawatt year is a million million watts of electricity made and used continuously throughout a year). The table shows what they found.

Fuel	Fatalities	Who	Deaths per twy
Coal	6400	Workers	342
Natural gas	1200	Workers and public	85
Hydro	4000	Public	883
Nuclear	31	Workers	8

Table 1. Deaths in the energy-producing industries, 1970–92.

I was astounded that they found nuclear energy to be the safest of all large-scale energy sources. The Swiss record puts it about forty times safer than taking energy by burning coal or oil and it is safer even than the renewable hydro-electricity. Yet so persistent have been the untruths about nuclear energy that we still regard taking energy from uranium in a reactor as more dangerous than burning carbon fuel in the oxygen of the air.

The persistent distortion of the truth about the health risks of nuclear energy should make us wonder if the other statements about nuclear energy are equally flawed. I wonder about the statement in August 2005 from the nuclear decommissioning authority, that it would cost £6 billion to decommission the UK's stocks of plutonium, as part of a £56 billion package for decommissioning the UK's nuclear installations. It is true that plutonium is a poisonous element and there is always a risk that it may be stolen to make nuclear weapons. But the stocks of plutonium in the UK have the energy equivalent of several hundred million tons of coal or oil, enough to keep the nuclear power stations of the UK running for several years. I find it incredible that our government and its advisers regard this abundant stock of nuclear fuel and our power stations as something to be decommissioned, written off; and they are prepared to pay over £60 billion to do it. Oil now costs $50 a barrel: at that price the UK stock of plutonium fuel alone is worth more than £100 billion in energy terms. It is all being done with stealth and pretence; we have never been asked if we were prepared to pay this huge cost.

Another flawed idea now circulating is that the world supply of uranium is so small that its use for energy would last only a few years. It is true that if the whole world chose to use uranium as its sole fuel, supplies of easily mined uranium would soon be exhausted. But there is a superabundance of low-grade uranium ore: most granite, for example, contains enough uranium to make its fuel capacity five times that of an equal mass of coal. India is already preparing to use its abundant supplies of thorium, an alternative nuclear fuel, in place of uranium.

THE RIGHT MIX OF ENERGY SOURCES

My strong pleas for nuclear energy come from a growing sense that we have little time left in which to install a reliable and secure supply of electricity; this is especially true in the United Kingdom and in several of the nations of Europe. I do not see nuclear energy as a panacea but as an essential part of a portfolio of energy sources. For the immediate future, and starting now, we need to exploit fission energy as much as we can as a temporary measure, while looking to a future when, having served our need, it can be replaced by clean energy from other sources. These should include renewables, fusion, and burning fossil fuel under conditions where the carbon dioxide effluent is safely sequestered, preferably in the form of an inert solid, such as magnesium carbonate. The important and overriding consideration is time; we have nuclear power now, and new nuclear building should be started immediately. All of the alternatives, including fusion energy, require decades of development before they can be employed on a scale that would significantly reduce emissions. In the next few years renewables will add an increment of emission-free energy, mainly from wind, but it is quite small when compared with the nuclear potential. Until 2008, when closures start, the UK nuclear generating capacity is 14,000 megawatts, and this is only 21 per cent of our total electricity production. To replace the nuclear output with one megawatt wind turbines would require 56,000 of them, and they would need to be backed up by a capacity of 10,500 megawatts of fossil fuel generators for those frequent occasions when the wind is too weak or too strong. Unless there are drastic changes in lifestyle we will have to go on using fossil fuel energy for several more decades; 30 per cent of our energy use is now for transport, and there is little chance that the carbon dioxide effluent of cars, trucks, trains and aircraft will be sequestered and buried.

The virtual superpower of Europe, Franco-Germany, has made the best of both worlds with its French half all nuclear and its German half all green This would be a fine and sensible solution were it not for Germany trying to make the rest of us support their industry by buying their wind turbines.

Meanwhile at the world's climate centres the barometer continues to fall and tell of the imminent danger of a climate storm whose severity the Earth has not endured for fifty-five million years. But in the cities the party goes on; how much longer before reality enters our minds?

6

Chemicals, Food and Raw Materials

At least 90 per cent of us in the first world now live in cities or in suburban areas around them. Even our holidays are usually spent in the urbanized resorts that have sprouted almost everywhere on Earth. Not only this, but few of us now walk through the countryside for enjoyment. Some of it is still beautiful, even though much is an agribusiness desert of monoculture fields fenced with barbed wire, or impassably muddy fields overstocked with cattle or sheep. This is not its natural state, and only those born before 1950 have seen how splendid it once was and could be again. Because our lives are so wholly urban, democracy ensures the election of governments almost entirely out of touch with the natural world.

Sandy and I often walk on the few remaining wild areas of the south-west region where we live. We particularly enjoy the coastal footpath that runs for 630 miles from Minehead in Somerset, passing Land's End on its way to Poole in Dorset. Even in high summer we encounter few other walkers, and most of these are within a hundred yards of the car parks along the trail, and they are often not so much walkers as users of the path as a lavatory for their dogs; yet so scenic is the section of path between Poole and Lyme Regis that it was chosen as a World Heritage Site and named the Jurassic Coast because the cliffs are sections through rocks that once formed the surface at the time of the great lizards. The fossils of this exciting period are displayed afresh on the beaches at Charmouth and Kimmeridge.

No wonder obesity is rife; we grow fat and die of metabolic diseases like diabetes, strokes and heart attacks, not just from overeating but as much from lack of exercise. Consider how few children ever walk to school or away from the city lights to see starlight sprinkled like jewels on black velvet or hear a cuckoo calling in the spring. Governments are aware that something could be wrong with the way we live and therefore have departments concerned with the environment. But a closer look reveals this deals mainly with roads and sewers and the planning of new towns. The natural world is acknowledged but mostly as 'unimproved' land suitable for wind farms, agribusiness, reservoirs and the other large-scale developments needed to service those living in the towns.

With such lifestyles and priorities it is no surprise that the natural world of Gaia seems foreign to many, who know little about the great Earth system that has for eons kept our planet a fit place for life. The only time we see the non-human world is vicariously on the television screen during wildlife programmes, or when astronauts share their vision of the Earth with us from space.

You would not subject yourself to surgery by a novice who had merely read books or viewed television documentaries on how to remove an inflamed appendix and had never before used a scalpel. Why should we trust urban environmentalists to advise our elected government on how to make laws intended for the welfare of our planet? Like amateur surgeons their intentions are good, but their execution is often woeful – or positively damaging. Of course, we still have compassion for wildlife and an aching nostalgia for a simpler more natural life. At any supermarket you will see consumers choose organic food, free-range eggs and food labelled as free of chemical contamination. Look at how we no longer use CFC-powered spray cans and we buy and run less-polluting cars. We really think we understand, and in some countries we even elect green political parties to play a role in government.

Yes, we do all of these things, and we mean well, but it is nowhere near enough and often the consequences are worse than inaction. This chapter looks at some of the blunders made in the name of environmentalism during the forty years since Rachel Carson's *Silent Spring* was published, showing what has happened for the good and where we have gone wrong.

CHEMICAL PESTICIDES AND HERBICIDES

Rachel Carson argued convincingly that the unregulated use of agricultural pesticides was leading to the widespread death of birds. She showed how birds eating insects poisoned by pesticides were damaged and was concerned that ultimately there would be such a death of birds that spring would be silent. In most parts of the world, the pesticides that bothered Carson have been banned from use, or strictly controlled in their application. Food grown and most meat slaughtered are checked for pesticide residues, and the legislation to control pesticides is working.

Carson pinpointed the abuse of chemical pesticides, and I suspect that innocent students, with the natural socialism of youth, imagine that DDT was devised by an employee of a monolithic chemical industry run by greedy capitalists solely for profit. In fact, the insecticidal properties of DDT were discovered by Professor Paul Herman Muller in 1939. He was rightly awarded the Nobel Prize for his discovery, which saved more lives than any other chemical previously invented. He was a good man and he generously gave the money from his Nobel Prize to his students, a most unusual gesture on the part of a professor. Yet Carson unintentionally made him a demon. It is important to remember the history of DDT; it was originally used against insect-borne disease, notably curbing the epidemic of typhus that ravaged Naples in the aftermath of the Second World War. Later it was used against mosquitoes, vectors of malaria, yellow fever and other tropical diseases. In this role it was, until it was banned, saving millions of lives yearly and vastly improving the quality of life of those hundreds of millions living in malarial regions; and in this use it was comparatively little threat to wildlife. DDT and other insecticides only became an environmental threat when agribusiness started using them on a large scale to improve crop yields. These insecticides badly needed controlling, but the indiscriminate banning of DDT and other chlorinated insecticides was a selfish, ill-informed act driven by affluent radicals in the first world. The inhabitants of tropical countries have paid a high price in death and illness as a result of their inability to use DDT as an effective controller of malaria.

I was more than a bystander as these events unfolded, first as the inventor of an extraordinarily sensitive instrument, the electron capture detector (ECD), that could detect quite infinitesimal traces of pesticides like DDT, and second as a science adviser to Lord Rothschild, then the coordinator of science for the company Shell, a major industrial maker of DDT, Dieldrin, and other chemical pesticides.

Lord Rothschild was also a distinguished biologist and a Fellow of the Royal Society. He and I were among the few scientists in the world at that time to have researched the biophysics of spermatozoa, and it was this topic that first brought us together. I shall not forget his pain and anger on reading Rachel Carson's book and experiencing the media storm that it provoked. As a naturalist he was agonized to discover the harm that his company had inadvertently caused, and he was enraged by the politicizing of what, in his opinion, could have been resolved in a seemly way.

We have to understand that the 'silent spring' did not come simply from poisoning by pesticides; the birds died because there was no longer space for them in our intensively farmed world. There are so many humans now aiming for a first-world lifestyle that we are displacing our partners on the planet, the other forms of life. We have to realize that cutting back emissions of greenhouse gases is only part of what we have to do; we have also to stop using the land surface as if it was ours alone. It is not: it belongs to the community of ecosystems that serve all life by regulating the climate and chemical composition of the Earth.

I make no apologies for repeating that Gaia is an evolutionary system in which any species, including humans, that persists with changes to the environment that lessen the survival of its progeny is doomed to extinction. By massively taking land to feed people and by fouling the air and water we are hampering Gaia's ability to regulate the Earth's climate and chemistry, and if we continue to do it we are in danger of extinction. We have in a sense stumbled into a war with Gaia, a war that we have no hope of winning. All that we can do is to make peace while we are still strong and not a broken rabble.

*

As someone who thought of himself as green I was alarmed by the new evidence of harm done by agricultural pesticides. Where I then lived in Wiltshire the whole landscape was being sterilized by enthusiastic young farm managers. Fast disappearing was the rich and biodiverse landscape of small meadows and hedgerows; replacing them were huge fields of monoculture barley and oilseed rape, and these were now all fenced with barbed wire. Bowerchalke, my village, had hardly changed since medieval times, with its five farms and those who worked on them. The young couples from village families expected to be able to rent or buy a cottage and continue their life there as their ancestors had done for hundreds of years. Within ten short years it had all changed; the farms were run by contract labour brought in from outside, the price of housing rose far beyond the villagers' ability to pay for it, and the village itself became an ex-urban colony peopled by the affluent middle class. This desecration of the rural scene, taking place across southern and eastern England, passed almost unnoticed, and few mourned the loss of biodiversity and of the village communities. This devastation continues and still escapes media attention; we could contrast it with the great outcry over the smaller-scale suffering and loss of the mining communities in the 1980s. Both events were disgraceful, but the lack of support and sympathy for the rural poor makes them an unclassified minority in our multicultural society. Bowerchalke had had a cricket team good enough to beat the County team of Somerset, a flourishing village school to which my children went and learnt the three Rs in the traditional way, and of course there was the village pub, The Bell, with its strict landlady, Chris Gulliver, who would allow no drunkenness or bad behaviour. Of course, few people actually suffered privation during the rape of rural life. The villagers were paid what were to them enormous sums for their cottages; the young who had no property were now moved to council houses and to new jobs in the nearby towns. What suffered were the birds, the animals and the wild plants of the countryside; fields with ancient hedgerows rich with colour in the spring and resounding with birdsong were now an empty expanse of monoculture grain. Urban folk also suffered the loss of a real countryside that could be wandered in and enjoyed as it had been in earlier times.

I knew a great deal about agribusiness and the reasons why those dogs of war had been let loose to destroy the English countryside. It was all in the interests of growing more food; we had nearly starved in the Second World War and were out to make efficient the old-fashioned and less productive English countryside. To picture what has happened, imagine a large garden with trees and shrubs and flower beds, and somewhere separate, a walled garden filled with kitchen vegetables; perhaps also a few sheep to keep the lawns cut. This was the countryside as it was before scientists found that such an inefficient use of land was wrong. The trees and shrubs must go, the lawns must be ploughed and planted with a single optimized crop suited to the local land and climate. This was what happened in England and much of Europe in the years from 1960 to 1980, and we felt dispossessed. The country I loved had all been taken away, and there was nothing that I could do personally to stop it. And it all had been done, as in every war, in the name of some ideology or other. A few greens seemed to share my burning hatred towards this new barbarism, and authors like Miriam Shoard, and Richard Mabey, especially in his recent book *Nature Cure*, have written of their concern. Sadly, many greens are now squarely behind a final solution to the problem of the rural regions: make them the place for industrial-scale renewable energy and let them be used for wind farming and for growing cash crops for bio fuels to keep the city lights glowing and the urban transport running. How can they talk of a green world with policies as black as this?

It is only human to be concerned for the welfare of fancy birds and cuddly animals living in Rousseau-style forests far away, but these are like the dandies of our own civilization, doing little of the hard work needed to keep Gaia going; that is done for the most part by the denizens of the soil, the micro organisms, the fungi, the worms, the slime moulds and the trees. Environmentalism has rarely been concerned with this natural proletariat, the underworld of nature; mostly it has been a radical political activity, and, not surprisingly, Rachel Carson's message was soon translated, at the dinner tables of the affluent suburbs and universities, from a threat to birds into a threat to people. In such a climate of opinion it was not long before scientists struggling for support found that research that seemed to suggest that compound X or pesticide Y was carcinogenic was unusually rewarding and brought fame and funds beyond their dreams. The media now had a near-endless source of stories and, later, courtroom dramas, as lawyers became involved in the compensation claims. The dinner-table talk was now intensified by fear, for nothing is as frightening in peacetime as the prospect of cancer. All chemicals were soon considered dangerous, and this gave respectability to the harmless and largely useless practices of alternative medicine. The desire for organic food, produced without man-made chemicals, became the inspiration of the greens. In other words, greens were drifting dangerously into an obsession with personal human problems. If we truly care about the welfare of mankind it is our duty to put Gaia first and our obligation to ensure that we do not take from her more than is our fair portion. To invoke Gaia without this in mind is no more than a counsel to perfection.

Fear of cancer in the first world led to blunt and unwise action against DDT and other similar chemicals without a proper consideration of the harm that might be done by denying those in the developing world the very real benefits that came from the sensible and proportionate use of DDT. The over-reaction against nitrates is another example of inappropriate legislation.

NITRATES

When we moved to our present home, Coombe Mill in Devon, nearly thirty years ago, the West Devon countryside was still idyllic, so different from what had become the agribusiness desert of our previous home in Wiltshire. A small river, the Carey, flows by Coombe Mill and is a tributary of the Tamar, whose course marks the boundary between Devon and Cornwall. In 1977 the River Carey was clear and sparkling and so rich with salmon and sea trout that water bailiffs patrolled it to prevent illegal fishing. Fishermen in waders occasionally strayed from the permitted fishing beats of the lower reaches of the river on to our land and told us their immemorial tales of the ones that got away.

Our region is one of the wetter parts of southern England, and heavy rains, especially in summer, make arable farming difficult. Most farmers around here, then and now, raise sheep and cattle fed by the rich abundant growth of grass. In 1977 they farmed as they always had done, by making hay in the late spring and early summer and storing it in stacks to provide food for their cattle in the winter. This low-key farming is what made England such a pleasant land to look at and to live in, and it provided a rich supply of food for the native wildlife. The pressure to grow more food that began during the Second World War, when our need was great, led to the spreading of nutrient chemicals on the fields; with more animals in the fields the manure they produced was not enough to complete the cycle of essential nutrient elements, particularly of nitrogen, which is essential for life. To make up the deficit, farmers spread nitrogen in the form of ammonium nitrate and potassium and phosphorus as potassium phosphate. A nitrate is a salt that comes from the combination of nitric acid with such alkalis as potassium, sodium or ammonium hydroxides; it is a white powder and is, like salt itself, soluble in water. Ammonium nitrate, the usual farm fertilizer, comes in huge plastic sacks containing hundreds of kilos of white granules. It is safe enough when used in farming, but terrorists have made their bombs from it. When nitrate, loses one of its oxygen atoms it becomes a new ion called nitrite. Nitrites are potentially dangerous because they readily react under acid conditions with amines, which are molecules with nitrogen atoms attached to two hydrogen atoms and any of a vast number of hydrocarbons. The products of the reaction are called nitrosamines. In 1963 I had met a medical scientist at a conference on radiation biology, W. Lijinsky. He had become famous through researching the carcinogenic properties of these nitrosamines. It was something of a shock to most of the older generation of chemists, for in their student days many of them had prepared tens or hundreds of grams of diethyl-nitrosamine as a student exercise. We wondered if breathing in the vapours of these unexpectedly poisonous compounds had set cancer timebombs ticking within our own bodies.

It was not long after this that concerned environmentalists discovered that nitrates naturally present in food and water supplies are changed in our saliva into nitrites, which are then imbibed daily with food and mixed with the acid of our stomachs. Amines are also naturally present in our food (they are what give fish its fishy smell), and these could react with the nitrous acid from the nitrites to form the potentially deadly nitrosamines. This information was used by activists who constantly inflamed concern about nitrates in food or drinking water until, in the 1970s, health authorities in Europe and the United States began to regard nitrate in food and water supplies as a dangerous threat to health. New strict regulations were then introduced to limit the use of nitrate as a fertilizer and to reduce its presence in food and water.

This new perception and legislation to limit the use of nitrates as fertilizers may have hastened the malign changes already at work in the countryside. The farmers of Devon and many other places were slowly changing the way they used grass, replacing haystacks with silos or plastic bags filled with silage. Within a few years they had stopped spreading ammonium nitrate on their fields and making hay. They adopted the modern procedure of slurry farming, where grass is gathered in springtime and converted into silage, a palatable – to cattle – dish not unlike pickled cabbage, or sauerkraut. This is a more efficient way of farming, and farmers benefited from storing grass as silage and could now stock more cattle on their land. Instead of using nitrate fertilizer, they now spread the dung collected in winter on their land, either directly or as slurry mixed with water. To an urban environmentalist, this was a proper organic way of farming. But by the early 1980s the pristine clear waters of the Carey had become brown and frothy and smelt just like an open sewer. In summertime, the quiet stretches where fish had risen to catch flies were covered with slimy green algae, flannel weed, and the river slowly died. The new organic slurry farming was loading the river with quantities of dung far beyond the amount it could digest. Every rainstorm washed dung from the fields into the river, and soon the oxygen level of the river fell to zero. Many of the partner species that make up a river ecosystem – the green plants that put oxygen in the water, the numerous insect species that live in the river and under the stones along its bed –

died, mainly from lack of light for photosynthesis and from anoxia. There was no longer any insect food for the river fish, so they had no chance of returning at times when slurry pollution was less. The problem would not have been so great had not farmers begun to feed their cattle in the winter on imported grain as well as silage; this allowed them to overstock with herds larger than the fields alone could support. As a consequence these fields that were once grass for the cattle to feed on now became, in addition, dumping grounds for the excess dung that had accumulated in the slurry pits and tanks during winter.

Over the years from 1977 until the mid eighties I was obliged to watch the river and the countryside die, and for me it was as moving an experience as any reported by Rachel Carson in *Silent Spring* on the death of birds. This time it was not the usual suspect, the chemical industry, that could be blamed, it was the fault of all of us and our misguided tendency to believe any accusation against big business. We all voted for the governments that passed the legislation for the control of nitrates while we turned a blind eye to the excesses of the European Common Agricultural Policy (CAP).

As always the real world is far more subtle and unpredictable than any of us think. The Carey now has a few fish in it, small trout and sticklebacks, but the burden of slurry still under the stones along the river bed will take decades to clear away and allow the river to come alive again. The improvement did not come from the dawning of wisdom but because my region was severely affected by two epizootics: foot and mouth disease and bovine spongiform encephalopathy. The cattle population plummeted.

Gaia is an intricately complex system and, in many ways, like our own bodies. It cannot be grossly manipulated to feed an ever-increasing burden of humans without consequences; all too often, panic driven by the fear of cancer leads to unwise and intemperate action.

A disturbing postscript to this tale of nitrates appeared in *Scientific American* in September 2004, which reported research that found nitrates in food and water are not harmful but beneficial to our bodies. We use them in the digestive process to assist the stomach acid to kill off pathogenic bacteria which all to often infest our food.

ACID RAIN

You will have gathered by now that many of the worst aspects of pollution are iatrogenic, that is, they arise from treatment that adds damage instead of curing the malady. Acid rain pollution provides an intriguing cautionary example of our unfortunate tendency to do harm while trying to do good.

Fred Pearce, in his 1987 book *Acid Rain*, provides a clear and readable account of the history of acid pollution. I didn't know until I read it that the Norwegian playwright Ibsen touched on the first symptoms of this malady of the industrial age. In one of his earlier plays, *Brand* (1866), he wrote:

> Direr visions, worse foreboding
> Glare upon me through the gloom,
> Britain's smoke-cloud sinks corroding
> On the land in noisome fume,
> Smilches all its tender bloom
> All its gracious verdure dashes
> Sweeping low with breadth of bane
> Steeling sunlight from the plain
> Showering down like rain of ashes.

One hundred years later, in the 1970s, the inhabitants of Norway and Sweden found to their dismay that the once abundant life of their lakes and rivers was declining, and chemical measurements strongly suggested that there was some change, or some pollution, that had made their waters too acid for life. Norway and Sweden are neither densely populated nor the site of as much heavy industry as the United Kingdom or Germany. So where was the acid coming from? It did not take them long to find the source. Water in the rain gauges of the meteorological stations of Scandinavia was even more acid than in the lakes and rivers. The destructive acid was being brought in as rain; so where was the rain coming from?

Everyone who lives in Western Europe knows that the prevailing wind is westerly, coming in from the Atlantic Ocean. The only sizeable landmass west, or rather south-west of the Nordic countries is the United Kingdom. It was well known that the UK produced most of its electricity by burning coal in huge power stations – the one at Drax in Yorkshire was the largest in the world. The acid-rain research was soon made public, and it became a major media interest in northern Europe. England was blamed as the main exporter of acid. The crime of exporting acid fitted everyone's prejudices – even the English themselves assisted, for we all knew that industry was evil and polluting and run only for profit (we conveniently forgot that the coal and electricity industries had been nationalized for more than twenty years). Everyone was sure that the English coal-burning power stations were to blame; that they were now run for the public good was no help at all.

In the 1980s, representatives of the national scientific academies of the Nordic nations and of England met to discuss the nature of the problem and likely remedies. It was not to be a trial where the accused is presumed innocent until proven guilty. My friends Sir John Mason and Sir Eric Denton, both Royal Society representatives at this meeting, tell me that the Nordic president in his opening address said, 'Gentlemen, we are here to prove that the British emissions of sulphur gases are the source of acid rain in Scandinavia.' There was little doubt about the verdict, and the United Kingdom accepted blame and agreed to place sulphur-removing devices on the smoke stacks of its major coal-burning power stations. Most people today in northern Europe believe that justice was done and the guilty culprits made to mend their polluting ways.

But there is more to acid rain than met Nordic eyes. As with Rachel Carson's story that pesticides from chemical industry were the sole agents destroying the birds and would lead eventually to a silent spring, it was not simply the British that were to blame. To be sure, the British coal-burning power stations were the source of some of the acid – about 17 per cent – falling on the Nordic lands. But on its own this would not be enough to cause the serious observed acidification of the Scandinavian rivers and lakes. So where on Earth was the larger part of the acid coming from?

	UK	USSR	Germany	Czech	Sweden	Norway
Sweden	6.8	12	14	4	18.5	2.4
Norway	15	8.3	15	3.5	4	10

Table 2. Proportions of sulphur (%) deposited in Sweden and Norway from other nations.

Table 2 illustrates the principal sources of acid precipitation (the data are taken from Bridgman's 1990 book). As much came from Germany as from the UK, and the eastern sector was then still part of the Soviet communist empire, a place where the good of the state came well before concern about pollution. Not only this, but East Germany burnt lignite in its power stations, a sulphur-rich coal that was abundant in that part of Europe. Some of the acid even came from the Nordic nations themselves.

Another source, most surprisingly, is the North Sea, though not to the extent I once thought. Microscopic algae living in the ocean make the gas DMS, which escapes to the air and there oxidizes to form sulphuric and methane sulphonic acids. In recent years the algae have done well from the nutrients in the flow of farming effluent that now contaminates the European rivers; both the Baltic and the North Sea are enriched with nutrients far above the level of the Atlantic Ocean. I shall not forget a visit to Schweningen, a resort on the sea coast of the Netherlands in 1990; here, as we walked on the beach, we were repelled by the mounds, feet high, of evil, sulphurous-smelling detritus that lay on the sand at the edge of the sea. It was an algal bloom, probably phyocystus, which the wind had blown onshore from the overfed North Sea. But in 1996 Sue Turner and her colleagues from the University of East Anglia published a paper that provided a comprehensive account of the natural emissions of DMS from the North Sea and the potential impact of these emissions on the European atmospheric sulphur budget. They found the annual emission of sulphur as DMS from the North Sea was only 0.4 per cent of the total industrial emissions of nations bordering the North Sea. On the other hand, algal emission is seasonal and can be local: Leck and Rodhe in 1991 estimated that in July the seas adjacent to Scandinavia emit between 0.8 and 3 times the amount of sulphur from Norwegian industrial sources. Nevertheless, I am glad to take this opportunity to correct the wrong impression given in my earlier book *Gaia: The Practical Science of Planetary Medicine* that natural emissions were a serious contributor to acid deposition in Scandinavia.

In 1988 I had the chance to ask the then head of the British power industry, Lord Marshall, why we had so meekly accepted all the blame for our sulphur emissions. Curtly, he replied, 'The cost of installing sulphur removers was trivial compared with those I then faced for the privatization of the electricity industry.' It is all too easy it seems to lose our sense of proportion.

This is not the end of the acid rain story. In response to the problem of acid rain, the EU introduced legislation to reduce the sulphur content of fuels as well as making sure that sulphur emissions from power stations were filtered out. The good physicians of Brussels were applying the therapy that we all thought was needed to cure the disease. Sadly, we now know that once again, iatrogenic illness can result. Recent research confirms what a few of us have long suspected, that the all-pervasive European atmospheric haze that blights the summer skies and reduces visibility, sometimes to no more than a few hundred yards, is a sulphate aerosol and a source of what is referred to as 'global dimming'. What we see is the acid of acid rain spread wide across the whole of Europe, even into Asia. Before you think that we must stop it, consider the advice of the scientists. They say that this haze is reflecting sunlight back to space and keeps those of us beneath it several degrees cooler than we might otherwise be. In some senses the acid rain aerosol is a partial cure for global warming. Just imagine how much worse the intense heat of summer in 2003 would have been without it, and how much worse it will be when this European legislation starts to work.

FOOD AS A HAZARD

Life in the city is starved of contact with the natural world, and I suspect that many imagine that plant life has somehow evolved so as to be our perfect food. Not long ago we were sure it had been created by a beneficent God solely for us to eat. It is surprising how few seem aware that plants dislike being eaten and will go to extraordinary lengths to deter, disable, or even kill any animal or invertebrate trying to eat them. Garlic may have a pleasing flavour for some of us, but in its evolution it has found that the synthesis of an odorous suite of sulphur compounds is an effective discouragement to most of the insects, animals and micro organisms in its environment. Try chewing an uncooked caper, the seed capsule of a plant of the euphorbia family, and you will be disabled by the pain and blisters in your mouth and on your lips. Yew and the castor oil plant go the whole way and will kill anyone, or any animal, foolish enough to chew instead of just swallowing their seeds.

The distinguished American physician Bruce Ames is famous for the Ames test, which detects the presence of any substance or radiation that changes an organism's genetic code. Code changes are called mutations, and these are usually fatal for the progeny of the damaged organism or at least lead to a diminished life; only rarely are mutations beneficial. Mutations can lead to cancer, and the agents of such mutations are called carcinogens. Certain naturally occurring substances are not carcinogenic in themselves but can cause mutated cells to become cancers, and these substances are called co-carcinogens. In a seminal article in *Science* in 1983, Ames described the ubiquity of carcinogens and co-carcinogens in the food we normally eat. Most importantly he revealed that natural carcinogens made by vegetable life were present at thousands of times higher abundances than were those from chemical industry. Those of us who make it a habit to eat only 'healthy' natural food should know that when we do we are imbibing an amazing variety of these natural substances that can make living cells malignant. If we are so unwise, or unlucky, as to eat nuts on which mould has grown, we could encounter one of the deadliest of carcinogens, aflatoxin. Despite this, the superstitious fear of 'man-made chemicals' is widespread, while natural chemicals are still regarded as beneficial. To carry this perverse urban illusion to its logical conclusion would make us believe that the poisons strychnine and botulinus toxin are harmless because they are natural. The wisdom of Paracelsus has been cast out, and no longer do we understand that even water is poisonous in excess and cyanide harmless in small doses. It is right to be cautious about the applications of science, as we often discover when scientific developments in medicine go wrong. But to embrace uncritically the untruths of new age medicine is foolish and dangerous, though the fact that the fit and young titillate their hypochondria with the harmless but limited practices of alternative medicine at least relieves the pressure on the overburdened health services. Less benign is the desire for organic food, produced without man-made chemicals. For me it is a monstrous irony, as my original training was as an 'organic' chemist, one skilled at making the chemicals that so many fear. The desire for organic food, food produced without the 'unnatural' addition of chemical fertilizers or pesticides, is a proper response to the excesses of agribusiness. But when I see the full shelves of organically grown food in the supermarkets, much of it imported from distant lands, I wonder if it is not an agribusiness of another kind. I see some of the proponents of organic food flying an anti-science flag while drifting dangerously into an obsession with personal human fears that ignores the real harm done to the Earth. As I have said before, we cannot farm more than about half the Earth's land surface without impairing Gaia's capacity to keep a comfortable planet. Sadly, at our present numbers the lower productivity of organic farms compared with intensive agriculture makes it a dubious enterprise. I am not alone in my criticism. Patrick Moore, a founder member of Greenpeace, shares my views. Because ultimately our welfare, even our survival, is wholly dependent on the health of Gaia, we ask the urban greens to think again and see that their primary obligation is to the living Earth. Humankind comes second.

PERCEPTION OF THE RISK

As we go about our daily lives we are almost all of us engaged in the demolition of Gaia. We do it every hour of every day, as we drive to work, shop or visit friends or as we fly to some distant holiday destination. We do it as we keep our homes and workplaces cool in summer or warm in winter. The sum total of all our pollutions has already added half a million million tons of carbon to the atmosphere; enough, if the geological records of the Eocene period fifty-five million years ago are correct and we continue to pollute, to start changing the world so completely that hardly any of our descendants will be there to see it. We will, by thinking selfishly only of the welfare of humans and ignoring Gaia, have caused our own near extinction.

That most esteemed scientist E. O. Wilson has repeatedly warned us, as have other distinguished biologists including Robert May and Norman Myers, that by taking natural habitats for agriculture we are causing an extinction of life comparable with that associated with the demise of the great lizards sixty-five million years ago. Their thoughts are confirmed by the Millennium Ecosystem Assessment of 2003, and by the 2005 report in *Science* by Jonathan Foley and his colleagues on the global consequences of land use. It is good that the life-science community shares with the Earth scientists of the IPCC a sense that we are in peril, but it is less helpful when they treat the threat as if it were wholly a biological one; they should have moved beyond the twentieth-century separation of the sciences. Perhaps it is too much to expect all scientists to speak with one voice in a common comprehensible language, but fortunately many of the climate community are beginning to do so. The scientists who form the IPCC and individual climatologists are well aware of the interconnectedness of the whole Earth system, including its life forms, and why this larger entity, and not merely the biosphere or an individual ecosystem, is important in the imminent climate change that will intensify extinction.

Despite all these warnings, we carry on destroying and seem to worry only about the nearly trivial, even imaginary, risk of cancer from mobile telephones, power lines, pesticide residues in food, or sunlight; topping them all is a fear of anything to do with nuclear energy. We are indeed straining at a gnat but swallowing a camel with ease.

Perhaps we know in our hearts the true nature of our peril and would rather face these minor imaginary risks than confront the ineluctable consequences of destruction. For many years now, sensible young men and women with their lives ahead of them have come to me to ask if there is any hope of a future for them. Such a question would never have occurred to me or my friends when we were young, even though the Second World War then loomed; we were confident of a rich and probably fulfilled life. Today it seems their intuitions, the unconscious summing up of the evidence coming into their sense about the world, give a gloomy message. In a similar way, perhaps, the stridency of the sceptics of global heating hides and reveals their fear that they may be wrong.

Fear of the Devil and of hell fire, so common in past centuries, now seems replaced by fear of cancer. Just as, in the past, fear was manipulated by the unscrupulous for personal gain, so there are now reincarnations of Iago, manipulating our natural fear and loathing of cancer for their own selfish agenda. Before we can counter their falsehoods we need to take a closer look at cancer and its causes.

If we survive the tragedy of global heating, historians will look back and see that one of our greatest errors was to be so frightened of cancer. The people of the first world have convinced themselves that chemicals and radiation stand in the way of their personal immortality. I was amazed to hear from an intelligent, middle-aged American woman the belief that the human lifespan was well over a hundred years; she had complete faith in the literal truth of the Old Testament and consequently felt that avoidable environmental poisons were cheating her of her natural life expectation. I suspect that this extraordinary delusion is quite common and is why so many do not realize that global change is a far greater threat to their lives.

What are the facts? About 30 per cent of us will die of cancer; few seem aware that the prime cause is breathing oxygen. One of the great ironies of Gaia's evolution is that animals are empowered by oxygen, which provides them with a huge gift of rapidly available energy – without it they would be as sessile as a tree – but the cost of this gift is a faster rate of death, and the price for Gaia is our ability to commit combustion.

Within each of the billions of cells that make up our bodies are tiny inclusions called mitochondria; these are the power stations of our cells. Inside these tiny particles, fuel from the food we have eaten reacts with the oxygen that we have breathed in. The output of energy from the mitochondria is a flood of molecule-sized rechargeable batteries, adenosine triphosphate (ATP) molecules, each able to power for an instant our muscles and our brains, so that we can walk and run and think. When discharged, these molecular batteries are recharged again at the mitochondrial power houses. For our bodies, with their billions of tiny mitochondria, the danger comes from the accidental leak of combustion products. As oxygen reacts with the food products, unintended pollutants are formed. These include the oxygen molecule with a negative charge called the superoxide ion, the hydroxyl radical and other highly reactive molecular species. These destructive molecules escape from the mitochondria as toxic pollutants and also arise accidentally anywhere in the body where oxygen can react unchecked. The omnipresence of oxygen in our bodies also greatly enhances the damage done by radiation and chemical poisons. The fiercely reactive radical products of oxidation will attack almost any other molecule they encounter, and this is how they damage the intricate orderly internal assembly of our cells. Almost all of this damage is repaired by an evolved set of enzymes and systems – which we could look on as the security services of oxygen-breathing life. But inevitably some damage is done to the genetic chemicals of our cells, like DNA, which are the programs and procedures for building new cells. Wonderfully, the damage to DNA is also repaired and there is a continuous check of its integrity.

In the course of a lifetime, unavoidably, a few of the billions of these comprehensive checks fail. From the failures to repair oxygen damage, new cells are born, with fatal or near-fatal disorders. Most of these damaged cells commit cellular suicide using a death pill that every cell possesses called a capsase. When this is activated it sets in course an orderly progression to dissolution. It is a miraculous process called apoptosis. Just imagine if each one of us, on concluding that he or she was so much more harmful than useful, began to take ourselves apart in so perfect a way that a tidy, orderly heap of spare parts for future human use was left.

Sometimes the damage done to DNA by the products of oxidation disables one of the genes that sets the instructions for cellular suicide, and when this happens a maverick cell is born and grows unchecked. Then, after several more potentially adverse changes, a fully unrestrained cancer cell is born. It grows and invades and eventually may kill the animal that spawned it.

This is no more than an imprecise sketch of carcinogenesis. We still lack knowledge of the finer details, but it is enough to show how the life-giving power of oxygen has a dark side. By the time we reach the biblical allotted span of seventy years, 30 per cent of us will have died of cancer, and for almost all of those deaths, breathing oxygen will have been the main cause.

Natural nuclear radiation coming from cosmic rays and from the radioactive elements in the soil, the air and our homes, can and does cause cancer, and it does so because it is energetic enough to split the abundant molecules of water in the living cell and liberate those same free radicals that come from oxidative metabolism. Other natural and man-made sources of cancer act like radiation, but none of them, apart from smoking cigarettes and too much sunburn, add significantly to the 30 per cent who die from breathing oxygen. Inflammation, as the name suggests, is a burning sensation and it is always accompanied by increased oxidation in the inflamed tissue and by an increased rate of cellular reproduction. Not surprisingly, it is associated with cancer. This is probably why some viral diseases, such as hepatitis B and C, cause cancer through chronic inflammation in the liver.*

Few of us are aware that the oxygen of the air is the dominant carcinogen of our environment, but multitudes are convinced by the untruth that most cancers are an avoidable consequence of environmental

* For those who want to know more, a balanced and comprehensible account is in Robert Weinberg's book *One Renegade Cell.*

pollution and there is an unceasing torrent of articles that sustain this false belief.

How on earth, you may ask, can something so good, something so benign as nuclear energy have been demonized to the point where people and sensible governments are frightened to use it? I think it is a consequence of the vulnerability of people to the astonishing power to deceive of an endlessly repeated falsehood. Advertising and propaganda and well-written fiction really do work, and most will continue to believe that 'nuclear' means 'deadly'. But you should occasionally ask why it is, in spite of us imbibing all that radioactivity and chemicals, the incidence of cancer has not perceptibly risen. And how is it that those who spend their working lives in nuclear power stations live longer than the general population, and far longer than coal miners? Because we are so frightened of cancer we tend to lose all sense of proportion. However much that fear may seem justified, there is no cause to be more fearful of it now; in spite of all our fears of cancer from radiation, from chemicals in food, and even from mobile phones and power lines, we live longer than ever.

I once lived in Houston, Texas, a wealthy American city which as a consequence has lawyers that are both competent and expensive. One of them, a famous trial lawyer, appeared on local television with an extraordinary offer. He invited anyone who was viewing and who had a murder in mind to go out and commit it, even if there were witnesses as unimpeachable as the Pope or the police commissioner. He then promised a defence that would ensure a verdict of not guilty at their trial; but, he added, it would cost them all that they possessed. His track record suggests that this was no idle boast. Now, I am not saying that the anti-nuclear movement or the CND-employed advocates are as powerful as the Houston lawyer, but they have succeeded in their aim of convincing the majority that anything nuclear is evil. To my mind this is as great a distortion of the truth and just as false as would have been the lawyer's advocacy on a trial jury. We are evolved to choose positively even when the choice may be wrong or irrational. When we choose a mate or buy a house, once the decision is made the choice is invested with virtue and those discarded are seen as loaded with disadvantages. This 'cognitive dissonance', now a curse as well as a blessing, can be summed up in the phrase, 'don't confuse me with facts, my mind is made up.'

7

Technology for a Sustainable Retreat

AMELIORATION

It seems likely that soon the United States will take global heating seriously and move from its recent scepticism. When they do, I believe that their response will be to try to stop it by a 'technological fix', the application of the skills they have acquired from their space programme and from their embracing of high technology.

There was a truly interesting scientific meeting at Cambridge University in January 2004, with the somewhat threatening title, 'Macro Engineering Options for Climate Change', which conjured in my mind visions of gigantic barriers spanning at least the Straits of Dover. Gathered at the Isaac Newton Institute in Cambridge was an unusual selection of scientists and engineers, nearly all of whom were concerned with global heating and planet-scale ideas for its amelioration.

The meeting was organized by Professor Harry Elderfield, a Cambridge Earth scientist, and Professor John Shepherd, an ocean scientist from Southampton University, and brought together the originators or advocates of a series of inspired ideas for stopping climate change by direct intervention at a planetary level. It was a serious meeting, and we were constrained from drifting into science fiction by informed and sensible critics in the audience. It soon emerged that there were two main approaches: the first to reduce the amount of heat received by the Earth from the sun, and the second to remove carbon dioxide or other greenhouse gases from the air or from combustion sources.

Direct and courageous answers to global warming were put forward by Lowell Wood and Ken Caldiera, from the Lawrence Livermore laboratory near San Francisco, who told us of their proposal to build in space a sunshade placed between the Earth and the sun. Wood described a sunlight-deflecting disc about seven miles in diameter placed at the Lagrange point in between the Earth and the sun (this is the point where the gravitational pull of the sun and of the Earth are equal and opposite and where little effort would be needed to keep the sun shade in place). He claimed that the disc could deflect or disperse a few percent of the incoming sunlight and so cool our planet. He made a persuasive argument that this unusual solution to global warming would be neither impossibly expensive nor impractical. It would weigh about 100 tons and could be assembled and spun out in space. He and Caldiera also put forward the possibility of using minute stratospheric balloons that would also reflect sunlight and achieve the same reduction of radiant heat from the sun.

An equally plausible way of lessening the input of solar radiation is to arrange the artificial production of marine stratus clouds across a large area of ocean surface; these are clouds or mist just above sea surface. John Latham, from the National Center for Atmospheric Research in Colorado, described small and practical devices that turn sea water into an aerosol of tiny particles that would serve to nucleate these clouds. This is a much more practical suggestion than it might at first sound: we already know that low-altitude marine stratus clouds form part of the natural cooling made possible by the emission by ocean algae of the gas dimethyl sulphide.

There was a general feeling that these procedures had potential, but Peter Liss, of the University of East Anglia, rightly pointed out that reducing the solar input solved only half of the problem; carbon dioxide coming from human activity would continue to increase in atmospheric abundance, and as it dissolved in the oceans their acidity would increase. There are good grounds for believing that acidity is detrimental to ocean productivity: early in 2005, Carol Turley and her colleagues at the Plymouth Marine Laboratory reported that the ocean had already become too acid for the comfort of marine organisms and that more carbon dioxide could be disastrous. The removal of carbon dioxide at the source or from the atmosphere itself was then discussed at length. From an engineering viewpoint it is entirely practical to remove carbon dioxide from smokestack gas, and it is not impossibly expensive to do so. The difficult problem with sequestering carbon dioxide is the vast volume of it and where to put it. An early solution tried was burial at sea; unfortunately the acidity problem already mentioned precludes this answer. It could be buried underground in used gas or oil fields; as mentioned before, this is already being done by the Norwegians in an exhausted gas field under the Norwegian ocean. Carbon dioxide could also be injected underground into appropriate rocks, but there is no certainty that such stores would be stable and that the gas would not sometimes be released. Natural topography could make such releases lethal, as revealed by a natural disaster in Cameroon a few years ago, where an escape of carbon dioxide from an extinct volcanic lake flowed as dense gas into a village and suffocated its inhabitants.

We seem blind to the dangers of the ever-increasing output of carbon dioxide; I feel the need to remind you that the yearly output of this gas would make a mountain one mile high and twelve miles in circumference. In August 2005, the Nuclear Decommissioning Authority (NDA) reported that nearly £60 billion would have to be spent decommissioning the nuclear installations of the UK in the next twenty-five years. It seems incredible that so large a sum should even be considered for so unimportant a task, when it would be much more worthwhile to spend it on ways to decommission the carbon dioxide. At the conference Ken Caldiera offered the practical suggestion that carbon dioxide be sequestered by reaction with a suspension of chalk in water. This would produce a solution of calcium bicarbonate that could be disposed of more easily than gaseous carbon dioxide.

We were intrigued by the American scientist Klaus Lackner, who proposed equipment to extract carbon dioxide directly from the air and then react it with a powder made from the alkaline igneous rock called serpentine. The resulting product would be magnesium carbonate, a stable solid that could in part be used as a building material and is easy to store compared with carbon dioxide itself. An attractive feature of his idea was that the process could be applied close to sources of serpentine rock and was not limited to sites near or at the sources of carbon dioxide.

Among the critics in the audience was the eminent economist Shimon Awerbuch, who wisely warned that anything we did to lessen the threat of global warming would only lead, while it lasted, to an even greater burning of fossil fuel, such is human nature.

We left the meeting with the feeling that although amelioration of global heating was a formidable problem it was not a hopeless prospect. I wondered if there might not be an even simpler way to cool the Earth. We might be able to imitate the well-known cooling effect of large volcanoes. Pinatubo, in the Philippines, when it erupted in 1991, injected sulphur dioxide into the stratosphere, where it oxidized to form an aerosol of sulphuric acid droplets. These droplets floated in the upper air for several years and offset significantly the greenhouse warming. We could put an aerosol of tiny sulphuric acid droplets in the stratosphere simply by requiring aircraft flying at that height to burn fuel containing a small amount of sulphur. The most heavily used air routes of the northern hemisphere are predominantly in the stratosphere. I discovered afterwards that this idea had already been proposed by the Russian scientist M. I. Budyko in the 1970s. It was rejected then on the grounds that it would encourage the over-consumption of fossil fuel. Now it might buy us the time needed to retreat sustainably.

Fuel suppliers normally *remove* sulphur-containing compounds from aviation fuel so as to reduce ground-level pollution. It would not be difficult to supply fuel containing between 0.1 and 1 per cent of sulphur, the amount needed for aerosol production. Of course, there would be problems such as those involved with the complex chemistry of stratospheric ozone depletion. Robert E. Dickinson, of the Arizona University Institute of Physics, has made a full and detailed study of amelioration by aerosols, and I recommend it to anyone interested in pursuing further this possible temporary escape from overheating.

As all too often, and because Gaia is still not part of our everyday thoughts, this excellent meeting in Cambridge failed to mention that climatology was only one part of global change. Just as important as curbing emissions is the need to recognize that the Earth's natural ecosystems regulate the climate and the chemistry of the Earth and are not there merely to supply us with food and raw materials. Our attempts to replace these ecosystems with farmland or forestry plantations have led in recent years, in Indonesia and elsewhere in the tropics, to disaster both regional and global in scale. In a *New Scientist* article in August 2005, Fred Pearce wrote of the ominous surface changes in Siberia and Alaska, where a recent 3°C rise in temperature has led to the widespread melting of frozen peat bog. He warned that this warming had the potential to release vast volumes of methane trapped by ice beneath the surface. I would add that, once the bog dries out, fires will add yet more carbon dioxide to the air: the overenthusiastic clearance of forest for agriculture in South East Asia, and the draining of the peat bogs in which the trees grew, has led to wild fires so massive that the output of carbon dioxide reached 40 per cent of the worldwide total from fossil fuel combustion. Less noticeable, but similarly destructive, are the long-term consequences of cattle and goat farming. I repeat the phrase 'Combustion, Cattle and Chainsaws are the three deadly Cs' – use them as little as possible.

I could not help thinking, after listening to this imaginative and thoughtful debate at Cambridge, that for anything that we do to change the surface or atmosphere of the Earth we need a constraining oath; something like the Hippocratic Oath that physicians utter: 'Do nothing that would harm the patient.' We need a warning placed on every bulldozer, chainsaw, and on all large energy-using devices: 'Do nothing that would harm the Earth.' As with the Hippocratic Oath it would be no more than a counsel to perfection, but far better than our present-day insensitive approach to the Earth's skin and atmosphere.

UTOPIAN FOOD AND LIFESTYLE

I suppose it might just be possible to support without disabling Gaia the eight billion people who will soon be living. To do it we would have to uncouple ourselves from the metabolism of the planet. We might, once fusion is running, produce all the energy we need, but we would still be farming far too much of the planetary surface, and no doubt threatening the ocean ecosystems also. So I like to speculate on the possibility that we could *synthesize* all the food needed by eight billion people, and thereby abandon agriculture. The global consumption of food is the equivalent of about 700 million tons of carbon each year, which is a small fraction of our current usage of carbon for fuel. The chemicals for food synthesis would come directly from the air, or more conveniently from carbon compounds sequestered from power station effluent. The nitrogen and sulphur could also come from these effluents, and all that we would need in addition would be water and trace elements. We would be acting like plants, but probably using fusion instead of solar energy.

What would be synthesized would not be the intricate, natural chemicals we now eat as broccoli, olives, apples, steaks or, more probably, hamburgers and pizzas. Rather, the large new food factories would make simple sugars and amino acids. This would be the feed stock for tissue cultures of meats and vegetables and for junk food made from any convenient organism that could be safely eaten. The technology would not be greatly different from that now employed in brewing beer or making antibiotics. By doing this on a scale large enough to feed everyone, the land now farmed could be released back to Gaia and used once again for its proper purpose, the regulation of the climate and chemistry of the Earth. The present over-fishing of the oceans could also cease.

I have also wondered if a small, densely populated nation such as Britain could be made viable and Gaia-friendly in the long term by dividing it into three parts. One third would be given for cities, industries, ports, airports and roads; the second third would be for intensive farming, enough to grow all we need; and the last third would be given entirely to Gaia and left to evolve wholly without interference or management.

Most of us prefer an urban existence, provided that predatory low-life is kept invisible. Dense, compact cities, free of suburban sprawl, the kind now favoured by the architect Richard Rogers in his book *Cities for a Small Planet* (1997); these would need comparatively little land and they might even be tight enough for walking to be the preferred method of transport. In a radio interview another distinguished architect, Norman Foster, reminded us that over 75 per cent of energy usage is in buildings and transport; dense, well-planned cities encourage its easy and painless reduction.

For longer-distance travel, to ease that peripatetic itch we all seem to have, we could use sailing ships again. I am not thinking of those magnificent wooden, four-masted vessels, whose operation required dozens of sailors. I imagine a high-tech automated sailing vessel, like a modern aircraft, that would travel a planned path chosen and updated to maximize the thrust of the wind. It would take longer than a jet but, as is often said, it is usually better to travel than to arrive. From the personal experience of thirteen transatlantic journeys on passenger ships to and from North America, it is far more pleasant to go by ship than by air, but, if air travel is demanded, then why not giant sailing airships that rode the trade winds? They could be made of aircraft materials and use steam as the lift gas.

We are, unconsciously, evolving to a state where much of our time is spent using low-energy devices. What a stunningly good invention was the mobile telephone: it exploits the universal tendency of humans to chatter and obliges us to consume hours of the day at minimal energy cost – it is one of the greenest inventions ever. Small computers of great efficiency are now stealing into our lives to make us spend our time at minimal energy cost, playing games or surfing the net. An ultra-high-tech low-energy civilization may well be possible, but it would be wholly different from the present-day vision of a low-energy world of sustainable development and renewable energy where the multitude tries to survive on food from organic small-holders farming a protesting Earth.

Whatever form future society takes it will be tribal, and hence there will be the privileged and the poor. This being so, there would in our high-tech world surely be a fashion among the rich for eating real food: vegetables grown in soil and cooked with meat and fish. We are in our present mess because the luxuries of whole-house heating and private transport by car have become necessities and far beyond the Earth's capacity to provide. Vigilance would be needed to constrain the growth of luxuries that threaten Gaia. I have to stress that the well being of Gaia must always come before that of ourselves: we cannot exist without Gaia.

8

A Personal View of Environmentalism

The concept of Gaia, a living planet, is for me the essential basis of a coherent and practical environmentalism; it counters the persistent belief that the Earth is a property, an estate, there to be exploited for the benefit of humankind. This false belief that we own the Earth, or are its stewards, allows us to pay lip service to environmental policies and programmes but to continue with business as usual. A glance at any financial newspaper confirms that our aim is still growth and development. We cheer at any new discovery of gas or oil deposits and regard the current rise in petroleum prices as a potential disaster, not a welcome curb on pollution. Few, even among climate scientists and ecologists, seem yet to realize fully the potential severity, or the imminence, of catastrophic global disaster; understanding is still in the conscious mind alone and not yet the visceral reaction of fear. We lack an intuitive sense, an instinct, that tells us when Gaia is in danger.

So how do we acquire, or reacquire, an instinct that recognizes not only the presence of the great Earth system but also its state of health? We do not have much to go on because the concepts of intuition and instinct tended to be ignored, or at best regarded as flaky and dubious, during the last two centuries of triumphant reductionism. In the twenty-first century we are somewhat freer to wonder about ideas like instinct and intuition, and it seems probable that long ago in our evolutionary history, when our ancestors were simple aquatic animals, we had already evolved an ability instantly to distinguish anything alive within the mainly inorganic ocean. This primeval instinct would have been supremely important for survival, since living things can be either edible, lovable or lethal. It is likely to be part of our genetic coding and hard wired into our brains so that we still have it in full strength. We do not need a doctorate in biology to distinguish a beetle from a stone, or a plum from a pebble. But, because of the circumscribed nature of its origins, the instinctive recognition of life is limited by the range of our senses and does not work for things smaller or larger than we can see. We recognize a paramecium as alive, but only when we can see it through a microscope. Even biologists, when they think of the biosphere, too often ignore all things smaller than can be seen with the naked eye. My friend and collaborator Lynn Margulis more than anyone has stressed the primary importance of micro organisms in Gaia, and she summarizes her thoughts in the book she wrote in 1986 with Dorian Sagan, *Microcosmos*. The Earth was never seen as a whole until astronauts viewed it for us from outside, and then we saw something very different from our expectation of a mere planet-sized ball of rock existing within a thin layer of air and water. Some astronauts, especially those who travelled as far as the moon, were deeply moved and saw the Earth itself as their home. Somehow we have to think like them and expand our instinctive recognition of life to include the Earth.

The ability instantly to recognize life, and other instincts, like the fear of heights and snakes, are part of our long evolutionary history, but there is another kind of instinct that is not innate but grows from childhood conditioning. The Jesuits discovered that a child's mind could be moulded to accept their faith, and that once done the child retained faith as an instinct throughout life; similar but different moulds fix lifelong tribal and national loyalty. The mind of a child is even plastic enough to be shaped to follow faithfully something as trivial as a football team or as potentially sinister as a political ideology. Abundant experience of this kind suggests that we could, if we chose, make Gaia an instinctive belief by exposing our children to the natural world, telling them how and why it is Gaia in action, and showing that they belong to it.

The founders of the great religions of Judaism, Christianity, Islam, Hinduism and Buddhism lived at times when we were far less numerous and lived in a way that was no burden to the Earth. Those holy men would have had no inkling of the troubled state of the planet a thousand or more years later, and their concern, rightly, would have been for human affairs. Rules and guidance were needed for individual, family and tribal good behaviour; we were the human family growing up in the natural world of Gaia and, like children, we took our home for granted and never questioned its existence. The success of these religious backgrounds is measured by their persistence as faiths and guides over more than a thousand years of further population expansion. When I was a child I was marinated in Christian belief, and still it unconsciously guides my thinking and behaviour. Now we face the consequences of fouling our planetary home, and new hazards loom that are much more difficult to understand or cope with than the tribal and personal conflicts of the past. Our religions have not yet given us the rules and guidance for our relationship with Gaia. The humanist concept of sustainable development and the Christian concept of stewardship are flawed by unconscious hubris. We have neither the knowledge nor the capacity to achieve them. We are no more qualified to be the stewards or developers of the Earth than are goats to be gardeners.

Perhaps Christians need a new Sermon on the Mount that sets out the human constraints needed for living decently with the Earth, and which spells out the rules for its achievement. I have long wished that the religions and the secular humanists might turn to the concept of Gaia and recognize that human rights and needs are not enough; those with faith could accept the Earth as part of God's creation and be troubled by its desecration. There are signs that church leaders are moving towards a theology of creation that could include Gaia. Rupert Shortt, in his book *God's Advocates* (2005), reported an interview with the Archbishop of Canterbury, Rowan Williams:

INTERVIEWER: The next question is that talk of miracles flies in the face of science. There is a lack of evidence for miracles as well as an intrinsic implausibility about them.

ARCHBISHOP: It is a very big issue, the question of divine action. Again, I think it has to be taken in connection with a doctrine of God rather than a very specific examination of any claim to start with.

> Let us put it this way. For a theological believer, the relation of God to creation is neither that of the old image of someone who winds up the watch and leaves it, nor is it that of a director in a theatre, or worse a puppet master who's constantly adjusting what's going on.
>
> It's the relation of an external activity which – moment by moment – energizes, makes real, makes active what there is. And I sometimes feel that a lot of our theology has lost that extraordinarily vivid or exhilarating sense of the world penetrated by divine energy in classical theological terms.

As I read on through these thoughtful and impressive responses I was taken back to the 1970s when Richard Dawkins and other strong-minded scientists fiercely contested the concept of Gaia using arguments similar to those they now use as atheists to challenge the concepts of God and creation. That argument with them about Gaia has I think been settled with an acceptance that Gaia is real to the extent that we have a self-regulating Earth but with a growing recognition that many natural phenomena are unknowable and can never be explained in classical reductionist terms – phenomena such as consciousness, life, the emergence of self-regulation and a growing list of happenings in the world of quantum physics. It is time, I think, that theologians shared with scientists their wonderful word, 'ineffable'; a word that expresses the thought that God is immanent but unknowable.

Important concepts like God or Gaia are not comprehensible in the limited space of our conscious minds, but they do have meaning in that inner part of our minds that is the seat of intuition. Our deep unconscious thoughts are not rationally constructed; they emerge fully formed as our conscience and an instinctive ability to distinguish good from evil. Perhaps this is why the early Quakers knew that the still, small voice within does not come from conscious reckoning. Our conscious rational minds are no more capable of deep thought than is the tiny screen of a contemporary mobile telephone able to present in its full glory a Vermeer painting. The extraordinary power of our unconscious minds is expressed in what we see as mundane things like walking, riding a bicycle or catching a ball. We would fail utterly to do any of these things by conscious thought; their automatic and instinctive achievement requires long and often tedious training. The same is true of inventors who, after long apprenticeship to their craft, become inspired to imagine and then construct devices that reveal emergence when they are switched on; physicists in a similar way exploit the incredible mysteries of quantum phenomena despite having no conscious understanding.

The history of science shows that we need to keep what is good in past interpretation of the world and merge in new knowledge as it appears. Newton's understanding enlightened physics for three hundred years. Einstein's relativity did not cast out Newtonian physics, it extended it. In a similar way, Darwin's great vision of evolution has raised biology from a cataloguing activity into a science, but now we are beginning to see Darwinism is incomplete. Evolution is not just a property of organisms – what evolves is the whole Earth system with its living and non-living parts existing as a tight coupled entity. It is foolish to think that we can explain science as it evolves, rationally and consciously. We have to use the crude tool of metaphor to translate conscious ideas into unconscious understanding. Just as the metaphor, a living Earth, used to explain Gaia, was wrongly rejected by reductionist scientists, so it may be wrong of them also to reject the metaphors and fables of the sacred texts. Crude they may be, but they serve to ignite an intuitive understanding of God and creation that cannot be falsified by rational argument.

As a scientist I know that Gaia theory is provisional and likely to be displaced by a larger and more complete view of the Earth. But for now I see it as the seed from which an instinctive environmentalism can grow; one that would instantly reveal planetary health or disease and help sustain a healthy world.

Green thoughts and ideas are as diverse and competitive as the plants of a forest ecosystem and, unlike the plants, they do not even share the spectral purity of the colour of chlorophyll. Green thoughts range from shades of red to shades of blue. The totalitarian greens, sometimes called ecofascists, would like to see most other humans eliminated in genocide and so leave a perfect Earth for them alone. At the other end of the spectrum are those who would like to see universal human welfare and rights, and somehow hope that luck, Gaia or sustainable development will allow this dream to come true. Greens could be defined as those who have sensed the deterioration of the natural world and would like to do something about it. They share a common environmentalism but differ greatly in the means for its achievement. Perhaps the most humane green arguments are in Jonathon Porritt's two books *Seeing Green* (1984) and *Playing Safe: Science and the Environment* (2000). He has done more than anyone I know to persuade the power bases of Europe to think and act in what he believes is an environmentally sound way, and he has selflessly devoted much of his life to this cause.

Since I met him at Dartington in 1982 I have thought of Jonathon as a friend, and therefore I deeply regret that in the past two years our paths have diverged; it is important that, deep though our differences are over the merits of nuclear and wind energy, we still share a great deal in common. In Chapters 5 and 6 I presented detailed criticisms of green thoughts and actions, but it was from within the environmental community, not from without, as in the recent book by Dick Taverne, *The March of Unreason* (2005), which expresses the viewpoint of an enlightenment liberal who rightly criticizes greens for their impractical romanticism. My feelings about modern environmentalism are more parallel with those that might pass through the mind of the headmistress of an inner-city school or the colonel of a newly formed regiment of licentious and naturally disobedient young men: how the hell can these unruly charges be disciplined and made effective?

The root of our problems with the environment comes from a lack of constraint on the growth of population. There is no single right number of people that we can have as a goal: the number varies with our way of life on the planet and the state of its health. It has varied naturally from a few million when we were hunters and gatherers to a fraction of a billion as simple farmers; but now it has grown to over six billion, which is wholly unsustainable in the present state of Gaia, even if we had the will and the ability to cut back.

If we could go back to, for example, 1840 and start again we might be able to reach a stable population of six billion if we were guided from the beginning by a proper understanding of the Earth. We would know that fossil-fuel combustion needed limiting and that cattle and sheep farming use far too much land and cannot be sustained, and that arable farming, with pigs and chickens as food animals consuming mainly vegetable waste, would be a better way to go. It might even be possible to sustain ten billion or more living in well-planned, dense cities and eating synthesized food.

If we can overcome the self-generated threat of deadly climate change, caused by our massive destruction of ecosystems and global pollution, our next task will be to ensure that our numbers are always commensurate with our and Gaia's capacity to nourish them. Personally I think we would be wise to aim at a stabilized population of about half to one billion, and then we would be free to live in many different ways without harming Gaia. At first this may seem a difficult, unpalatable, even hopeless task, but events during the last century suggest that it might be easier than we think. Thus in prosperous societies, when women are given a fair chance to develop their potential they choose voluntarily to be less fecund. It is only a small step towards a better way of living with Gaia, and it has brought with it problems of a distorted age structure in society and dysfunctional family life, but it is a seed of optimism from which other voluntary controls could grow and surely far better than the cold concept of eugenics that withered in its own amorality. In the end, as always, Gaia will do the culling and eliminate those that break her rules. We have the choice to accept this fate or plan our own destiny within Gaia. Whatever we choose to do we have always to ask, what are the consequences?

The regulation of fecundity is part of population control, but the regulation of the death rate is also important. Here, too, people in affluent societies are choosing voluntarily seemly ways to die. Traditionally, hospitals have for the elderly been places for dying in comparative comfort and painlessness; the hospice movement has served to set standards and make this otherwise unmentionable role of the health systems acceptable. According to Hodkinson, in his book *An Outline of Geriatrics*, about 25 per cent of the elderly entering hospitals die within two months. Now that the Earth is in imminent danger of a transition to a hot and inhospitable state, it seems amoral to strive ostentatiously to extend our personal lifespan beyond its normal biological limit of about one hundred years. When I was a young postdoctoral fellow at Harvard Medical School in Boston an eminent paediatrician complained of the huge, more than tenfold, disparity between funds given for cancer research and those given for childhood disease; I suspect that it still exists.

We have severed nearly all the natural physical constraints on the growth of our species: we can live anywhere from the Arctic to the tropics and, while they last, our water supplies are piped to us; our only significant predator now is the occasional micro organism that briefly mounts a pandemic. If we are to continue as a civilization that successfully avoids natural catastrophes, we have to make our own constraints on growth and make them strong and make them now.

Over half the Earth's people live in cities, and they hardly ever see, feel or hear the natural world. Therefore our first duty if we are green should be to convince them that the real world is the living Earth and that they and their city lives are a part of it and wholly dependent on it for their existence. Our role is to teach and to set an example by our lives. In purely human affairs, Gandhi showed how to do it; his modern equivalents might come from the Deep Ecology movement, founded by the Norwegian philosopher Arne Naess. I am moved by the ideas of deep ecology and touch on them again in the next chapter. In certain ways my long-time friend Edward Goldsmith is one of the few who have tried to live and think as a deep ecologist. His erudite and thought-provoking book *The Way* is essential reading for anyone who wants to know more about green philosophy; he founded *The Ecologist*, a magazine concerned with green thoughts and politics. It is now managed in much the same way by his nephew, Zac Goldsmith. The difference between us lies in our origins. I, not surprisingly, since my first experience in science was twenty-three years of medical research, think like a physician or even a surgeon. This is why I would like to see us use our technical skills to cure the ills of the Earth as well as those of humans. Teddy Goldsmith and the deep ecologists, from their humanist origins, scorn modern technology and would prefer alternative technology and medicine and would let Nature take its course. I acknowledge that they may be right and that iatrogenic illness, the disease that treatment causes, is all too common, but I cannot stand aside while civilization drinks itself to death on fossil fuels. And this is why I regard nuclear energy, however much it is feared, as a needed remedy.

The green community should have been reluctant to found lobbies and political parties; both are concerned with people and their problems, and, like megaphones, they amplify the demagogic voices of their leaders. Our task as individuals is to think of Gaia first. In no way does this make us inhuman or uncaring; our survival as a species is wholly dependent on Gaia and on our acceptance of her discipline.

I am often asked, 'What is our place in Gaia?' To answer we need to look back a long time ago in human history to when we were an animal, a primate, living within Gaia and different from other species only in unimportant ways. Our role then was like theirs, to recycle carbon and other elements. We lived on an omnivorous diet and returned to the air as carbon dioxide the carbon collected in their lifetimes by our food animals and plants. We had our niche in the evolutionary system, and our numbers were probably not more than a million.

As intelligent predators, we were equipped with useful brains and hands and could alter the boundaries of our niche in ways that were unavailable to other animals. We could throw stones, use simple stone and wood tools, and do it better than other primates.

Many animals, even insects like bees and ants, can communicate. They use alarms and mating calls and pass on detailed information about the size, direction and distance of food sources. We humans were fortunate to acquire through a mutation the ability to modulate our voices sufficiently for a primitive spoken language. This change was as profound for us as primitive people as the invention of the computer or mobile phone has been for modern humans. The members of the tribe could share experiences; they could plan ahead against drought and famine and guard against predators. We were by then the emerging Homo sapiens and may have been the first animals consciously to modify the environment for their own benefit. Most remarkably, we used natural fires started by lightning for cooking, clearing land and hunting.

The innocent among the urban intelligentsia think and talk of early humans as living in harmony with the natural world. Some of them go further and gather funds to preserve what they see as natural communities living in remote regions, such as the tropical forests. They see the modern world as clever but bad and these simple lifestyles as natural and good. They are wrong. We should not think of early humans as better or worse than we are; indeed, they were probably very little different.

Others consider us superior because of our cultured ways and intellectual tendencies; our technology lets us drive cars, use word processors and travel great distances by air. Some of us live in air-conditioned houses and we are entertained by the media. We think that we are more intelligent than stone-agers, yet how many modern humans could live successfully in caves, or would know how to light wood fires for cooking, or make clothes and shoes from animal skins or bows and arrows good enough to keep their families fed? I am indebted to Jerry Glynn and Theodore Gray for making this point in their guidebook for users of the computer program *Mathematica*, a mathematics processor. Using as an example the fact that modern children can hardly add a column of numbers without a calculator, they observe that this is no bad thing, since each stage of human development brings with it a full measure of skills exchanged for others no longer needed; stone-agers were probably as fully occupied with living as we are.

One group of these early humans migrated to Australia at a time when the sea levels were much lower than now and the journey by boat or raft was probably neither long nor difficult. From this group are descended the modern Australian aboriginals, often claimed to be an example of natural humans at peace with the Earth. Yet their method of clearing forests by fire may have destroyed the natural forests of the Australian continent as surely as do modern men with chainsaws. Peace on you Aboriginals; you individually are no worse and no better than we are, it is just that we are power-assisted and more numerous.

Through Gaia I see science and technology as traits possessed by humans that have the potential for great good and great harm. Because we are part of, and not separate from Gaia, our intelligence is a new capacity and strength for her as well as a new danger. Evolution is iterative, mistakes are made, blunders committed; but in time that great eraser and corrector, natural selection, usually keeps a neat and tidy world. Perhaps our and Gaia's greatest error was the conscious abuse of fire. Cooking meat over a wood fire may have been acceptable, but the deliberate destruction of whole ecosystems by fire merely to drive out the animals within was surely our first great sin against the living Earth. It has haunted us ever since and combustion could now be our *auto de fé*, and the cause of our extinction.

9
Beyond the Terminus

Like the Norns in Wagner's *Der Ring des Nibelungen*, we are at the end of our tether, and the rope, whose weave defines our fate, is about to break.

Gaia, the living Earth, is old and not as strong as she was two billion years ago. She struggles to keep the Earth cool enough for her myriad forms of life against the ineluctable increase of the sun's heat. But to add to her difficulties, one of those forms of life, humans, disputatious tribal animals with dreams of conquest even of other planets, has tried to rule the Earth for their own benefit alone. With breathtaking insolence they have taken the stores of carbon that Gaia buried to keep oxygen at its proper level and burnt them. In so doing they have usurped Gaia's authority and thwarted her obligation to keep the planet fit for life; they thought only of their own comfort and convenience.

Some time towards the end of the 1960s I walked along the quiet back lane of Bowerchalke village with my friend and near neighbour William Golding; we were talking about a recent visit I had made to the Jet Propulsion Laboratory in California and the idea of searching for life on other planets. I told him why I thought that both Mars and Venus were lifeless and that the Earth was more than just a planet with life, and why I saw it somehow in certain ways alive. He immediately said, 'If you intend to put forward so large an idea you must give it a proper name, and I suggest that you call it Gaia.' I was truly grateful to have his gift of this simple, powerful name for my ideas about the Earth. I gladly accepted it then as a scientist acknowledging an earlier literary reference, just as others in previous centuries referred to Gaia when naming the Earth sciences geology, geography and so on. At that time I knew little of Gaia's biography as a Greek goddess and never imagined that the New Age, then just beginning, would take Gaia as a mythic goddess again. In a way, however harmful this has been to the acceptance of the theory in science, the New Agers were more prescient than the scientists. We now see that the great Earth system, Gaia, behaves like the other mythic goddesses, Khali and Nemesis; she acts as a mother who is nurturing but ruthlessly cruel towards transgressors, even when they are her progeny.

I know that to personalize the Earth System as Gaia, as I have often done and continue to do in this book, irritates the scientifically correct, but I am unrepentant because metaphors are more than ever needed for a widespread comprehension of the true nature of the Earth and an understanding of the lethal dangers that lie ahead.

After forty years living with the concept of Gaia I thought I knew her, but I realize now that I underestimated the severity of her discipline. I knew that our self-regulating Earth had evolved from those organisms that left a better environment for their progeny and by the elimination of those who fouled their habitat, but I never realized just how destructive we were, or that we had so grievously damaged the Earth that Gaia now threatens us with the ultimate punishment of extinction.

I am not a pessimist and have always imagined that good in the end would prevail. When our Astronomer Royal, Sir Martin Rees, now President of the Royal Society, published in 2004 his book *Our Final Century*, he dared to think and write about the end of civilization and the human race. I enjoyed it as a good read, full of wisdom, but took it as no more than a speculation among friends and nothing to lose sleep over.

I was so wrong; it was prescient, for now the evidence coming in from the watchers around the world brings news of an imminent shift in our climate towards one that could easily be described as Hell: so hot, so deadly that only a handful of the teeming billions now alive will survive. We have made this appalling mess of the planet and mostly with rampant liberal good intentions. Even now, when the bell has started tolling to mark our ending, we still talk of sustainable development and renewable energy as if these feeble offerings would be accepted by Gaia as an appropriate and affordable sacrifice. We are like a careless and thoughtless family member whose presence is destructive and who seems to think that an apology is enough. We are part of the Gaian family, and valued as such, but until we stop acting as if human welfare was all that mattered, and was the excuse for our bad behaviour, all talk of further development of any kind is unacceptable.

So often when disaster visits we still cry, 'How could God have let this happen?' And now that there is a probability that most of us will perish, can belief in God continue? Darwin once described the evolutionary process as 'clumsy, wasteful, blundering, low and horribly cruel'. But surely not as cruel, or as culpable, as we have been and still are to the rest of life on Earth; especially since so many other innocent organisms will share our fate.

It would be easy to think of ourselves and our families as incarcerated in a planet-sized condemned cell – a cosmic death row – awaiting inevitable execution. The days and years will pass, the seasons continue and we will be fed and entertained, and if we have faith we will ask God for a reprieve. Some like Sandy and me will probably cheat the executioner and die before our time is due; the cruel consequences will come for our children and grandchildren.

I am a scientist and think in terms of probabilities not certainties and so I am an agnostic. But there is a deep need in all of us for trust in something larger than ourselves, and I put my trust in Gaia, and declared it in my autobiography, *Homage to Gaia*, in 2000. Was ever a trust so severely tested?

As is often the way with lesser crises I turn to my friend and mentor, Sir Crispin Tickell, and it happened that he had an answer in the form of an address he gave before a conference on The Earth Our Destiny, at Portsmouth Cathedral in 2002. It was a deeply moving, wise and helpful observation on our place in the environment. The last paragraphs of the text were:

The ideology of industrial society, driven by notions about economic growth, ever-rising standards of living, and faith in the technological fix, is in the long run unworkable. In changing our ideas, we have to look forward towards the eventual target of a human society in which population, use of resources, disposal of waste, and environment are generally in healthy balance.

Above all we have to look at life with respect and wonder. We need an ethical system in which the natural world has value not just for human welfare but for and in itself. The universe is something internal as well as external.

He concluded with the words of the twelfth-century abbess Hildegard of Bingen, who wrote of God:

> . . . I ignite the beauty of the plains,
> I sparkle the waters,
> I burn in the sun, and the moon and the stars . . .
> I adorn all of the Earth,
> I am the breeze that nurtures all things green . . .
> I am the rain coming from the dew that causes the grasses to laugh with the joy of life.
>
> Let us likewise rejoice.

In certain ways the human world is re-enacting the tragedy of Napoleon's advance on Moscow in 1812. In September of that year, when he reached the Russian capital, he had already gone too far, and his precious supplies were daily being consumed while he consolidated his capture. He was unaware that the irresistible forces commanded by General Winter were siding with the Russians, allowing them to counter-attack and regain their losses. The only way he could have avoided defeat was an immediate and professionally executed retreat so that his army could remain intact to fight another time. The quality of generalship is measured in military circles by the ability to carry through and organize a successful retreat.

The British remember with pride the successful withdrawal of their army from Dunkirk in 1940, and do not see it as an ignominious defeat. It was certainly not a victory, but it was a successful and sustainable retreat. The time has come when all of us must plan a retreat from the unsustainable place that we have now reached through the inappropriate use of technology; far better to withdraw now while we still have the energy and the time. Like Napoleon in Moscow we have too many mouths to feed and resources that diminish daily while we make up our minds. The retreat from Dunkirk was not just good generalship: it was aided by an amazing expression of spontaneous unselfish good will from those numerous civilians who willingly risked their lives and their small boats to cross the channel to rescue their army. We need the people of the world to sense the real and present danger so that they will spontaneously mobilize and unstintingly bring about an orderly and sustainable withdrawal to a world where we try to live in harmony with Gaia.

Economists and politicians have to square the utter necessity of a rapid and controlled shutdown of emissions from fossil fuel burning with the human needs of civilization. Economic growth is as addictive to the body politic as is heroin to one of us; perhaps we have to keep the craving in check by using a safer substitute, an economist's methadone. I would suggest again that the mobile phone, the internet and entertainment from computers are moves in the right direction; they use time and energy that might otherwise be spent travelling by car or aircraft. Moreover, there is information technology and the efficient use of energy, for example using the ultra efficient white light emitting diodes (WLEDs) to see at night. Should technology of this kind become the main source of economic growth it would let us spend our lives harmlessly and fill some of the time that now we use in fuel-consuming travel. To an extent we are evolving that way.

Until quite recently, although many of us were aware that serious environmental change could happen and believed the predictions of the IPCC, somehow our knowledge seemed theoretical and academic, not indicating that something deadly was imminent. It was a small event that awakened me to these dangers. Fear crystallized as sharp needles in the supersaturated spaces of my mind when, in October 2003, my near neighbours, Christine and Peter Hadden, told me of plans to erect giant wind turbines in the countryside near our homes. Suddenly I realized what our politicians meant by sustainable development and renewable energy, and what it would do to the last remaining good countryside of West Devon. I could almost hear them say, 'Let us harvest the wind for energy, and plant bio fuel crops to keep the cars of urban voters running. We can do it without polluting the air or tangling with that nasty, dirty, fearful nuclear stuff.'

By good countryside I mean farming land and communities that live well with the Earth and represent an ecosystem which, although dominated by people, has ample room left for woodlands, hedgerows and meadows. Most of southern England was like this before 1940, and the largest remaining parts are in the West Country, especially in Devon. In my mind these last remaining areas of countryside were the face of Gaia, and it was about to be sacrificed. It was this that awakened my fury, and made me fully aware of the coming crisis of global heating. To make good countryside into industrial parks for wind energy merely as a gesture to prove their environmental credentials showed how far our leaders were from understanding our peril. To keep their urban enclaves comfortable, they would devastate by industrial development the remaining areas of good countryside.

I moved to West Devon twenty-eight years ago to escape the bulldozers that were destroying the Wiltshire hedgerows and meadows. Unwisely I thought that the gentle farmland of Devon was too poor to be developed and would let me live out my life in a countryside I loved. I had not allowed for incessant ideological good intentions and the near-religious belief in renewable energy and sustainable development for the good of us all.

They call Sandy and me 'NIMBYs' because we fight their final solution to the energy problem. Perhaps we are NIMBYs, but we see those urban politicians as like some unthinking physicians who have forgotten their Hippocratic Oath and are trying to keep alive a dying civilization by useless and inappropriate chemotherapy when there is no hope of cure and the treatment renders the last stages of life unbearable.

So is our civilization doomed, and will this century mark its end with a massive decline in population, leaving an impoverished few survivors in a torrid society ruled by warlords on a hostile and disabled planet? I hope that it will not be that bad; once a technically advanced nation wakes up to its responsibility, perhaps in response to our alarm call, they will say 'we can fix it.' They might use something like space-mounted sunshades or Latham's floating nuclei generators that put white reflecting clouds across the ocean surface. Technological fix it may be, but if it works we have only ourselves to blame if we do not take advantage.

Sunshades for cooling the Earth are more valuable than they might at first appear; they could wholly neutralize the harmful effects of unscheduled methane releases. They might even provide an adjustable remedy ready to offset the global heating should the methane clathrates of the ocean suddenly escape into the atmosphere. Keeping in mind the similarity of the Earth's physiology to that of a human, it is useful to compare such a technological fix with the use by paramedics of oxygen for heart failure and breathing difficulty, or a pressure pad for haemorrhage – something temporary, to keep a patient alive until they reach the full services of a hospital.

By itself this fix will do no more than buy us time to change our damaging way of life, because if we continue to burn fossil fuels and let the carbon dioxide rise in abundance, ocean life, essential to the health of Gaia, will be further damaged. But we may risk it because time is needed to instal equipment for carbon sequestration and for nuclear fusion and whatever forms of economically sensible renewable energy become available. In the longer term we have to understand that however benign a technological solution may seem it has the potential to set humanity on a path to the ultimate form of slavery. The more we meddle with the Earth's composition and try to fix its climate, the more we take on the responsibility for keeping the Earth a fit place for life, until eventually our whole lives may be spent in drudgery doing the tasks that previously Gaia had freely done for over three billion years. This would be the worst of fates for us and reduce us to a truly miserable state, where we were forever wondering whether anyone, any nation or any international body could be trusted to regulate the climate and the atmospheric composition. The idea that humans are yet intelligent enough to serve as stewards of the Earth is among the most hubristic ever.

So what should a sensible European government be doing now? I think we have little option but to prepare for the worst and assume that we have already passed the threshold. Like paramedics, their first priority is to keep the patient, civilization, alive during the journey to a world that at least is no longer undergoing rapid change. We face unrestrained heat, and its consequences will be with us within no more than a few decades. We should now be preparing for a rise of sea level, spells of near-intolerable heat like that in Central Europe in 2003, and storms of unprecedented severity. We should also be prepared for surprises, deadly local or regional events that are wholly unpredictable. The immediate need is secure and safe sources of energy to keep the lights of civilization burning and for the preparation of our defences against the rising sea level. There is no alternative but nuclear fission energy until fusion energy and sensible forms of renewable energy arrive as a truly long-term provider. Nuclear energy is free of emissions and independent of imports from what will be a disturbed world. We would be right to cut back all emissions to a minimum, and this includes emissions of methane from leaking pipes and landfill sites. But most of all we need electricity to sustain our technologically based civilization.

In several ways we are unintentionally at war with Gaia, and to survive with our civilization intact we urgently need to make a just peace with Gaia while we are strong enough to negotiate and not a defeated, broken rabble on the way to extinction. Can the present-day democracies, with their noisy media and special-interest lobbies, act fast enough for an effective defence against Gaia? We may need restrictions, rationing and the call to service that were familiar in wartime and in addition suffer for a while a loss of freedom. We will need a small permanent group of strategists who, as in wartime, will try to out-think our Earthly enemy and be ready for the surprises bound to come. Globally, the climate agencies of the UN have performed magnificently, as the IPCC proves. But as the climate worsens individual nations will need more and more to address disasters locally as they happen. In a sense, the great party of the twentieth century, with its extravagant overspending and its war games, is over. Now is the time for washing up and throwing out the debris.

My wisest of friends, Jane and Peter Horton, have warned me that the metaphor of war and battles with Gaia is masculine and could be offensive to women who now at last have power and influence on the way we act. They prefer my metaphor of Gaia as the stern but nurturing mother. They may well be right, but I ask them, as I ask Earth scientists who so dislike my image of a living Earth, to consider metaphor seriously as a path to the primitive feelings of the unconscious part of our minds. We are two sexes who respond differently and both metaphors may be needed. We belong to the family of Gaia and are like a revolting teenager, intelligent and with great potential, but far too greedy and selfish for our own good.

Men and women both need to be aware of what we are missing. Already for most of us the artificial world of the city is the whole of our lives and we think that to survive all we need is to be streetwise. But even in the city a few remnants of the natural world still continue in the parks and gardens. Make the most of them, for they continue to die away, as does the countryside many know and love; they are precious indeed.

If it should be that we have already passed the threshold of irreversible heating, then perhaps we should listen to the deep ecologists and let them be our guide. One of them that I know well as a friend is the biologist Stephan Harding, and I am indebted to him for making me aware of deep ecology. This small band of deep ecologists seem to realize more than other green thinkers the magnitude of the change of mind needed to bring us back to peace within Gaia, the living Earth. Like the holy men and women who make their whole lives a testament to their faith, the deep ecologists try to live as a Gaian example for us all to follow.

Few of us now can change our lives sufficiently to express our allegiance to Gaia as they do, but I suspect the changes soon to come will force the pace, and just as civilization ultimately benefited in the earlier dark ages from the example of those with faith in God, so we might benefit from those brave deep ecologists with trust in Gaia. The monasteries carried through that earlier Dark Age the hard-won knowledge of the Greek and Roman civilizations, and perhaps these present-day guardians could do the same for us. Despite all our efforts to retreat sustainably, we may be unable to prevent a global decline into a chaotic world ruled by brutal war lords on a devastated Earth. If this happens we should think of those small groups of monks in mountain fastnesses like Montserrat or on islands like Iona and Lindisfarne who served this vital purpose.

Few travellers from the north would go to the tropical south without antimalarial drugs, or to the Middle East without checking how the local war was progressing. By comparison our journey into the future is amazingly unprepared. Where people know well the local danger, as in Tokyo, they prepare for the earthquake to come. When the threats are global in scale we ignore them. Volcanoes, like Tamboura, Indonesia, in 1814 and Laki, Iceland, in 1783, were much more powerful than was Pinatubo in the Philippines (1991), or Krakatoa in Indonesia (1887). They affected the climate enough to cause famine, even when our numbers were only a tenth of what they are now. Should one of these volcanoes stage a repeat performance, do we have now enough stored food for tomorrow's multitudes? If part of the Greenland or Southern glaciers slid into the sea, the level of the sea might rise by a metre all over the world. This event would render homeless millions of those living in coastal cities. Citizens would suddenly become refugees. Do we have the food and shelter needed when cities such as London, Calcutta, Miami and Rotterdam become uninhabitable?

We are sensible and we do not agonize over these possible doom scenarios. We prefer to assume that they will not happen in our lifetimes. We take them no more seriously than our forefathers took the prospect of Hell, but the thought of appearing foolish still scares us. An old verse goes, 'They thieve and plot and toil and plod and go to church on Sunday. It's true enough that some fear God but they all fear Mrs Grundy.' In science we have our Drs Grundy also, and they are all too eager to scorn any departure from the perceived dogma. Scientists and science advisers are afraid to admit that sometimes they do not know what will happen. They are cautious about their predictions and do not care to speak in a way that might threaten business as usual. This tendency leaves us unprepared for a catastrophe such as a global event that is wholly unexpected and unpredicted – something like the creation of the ozone hole but much more serious; something that could throw us into a new dark age.

We can neither prepare against all possibilities, nor easily change our ways enough to stop breeding and polluting. Those who believe in the precautionary principle would have us give up, or greatly decrease, burning fossil fuel. They warn that the carbon dioxide byproduct of this energy source may sooner or later change, or even destabilize, the climate. Most of us know in our hearts that these warnings should be heeded but know not what to do about it. Few of us will reduce their personal use of fossil fuel energy to warm, or cool, their homes or drive their cars. We suspect that we should not wait to act until there is visible evidence of malign climate change – for by then it might be too late to reverse the changes we have set in motion. We are like the smoker who enjoys a cigarette and imagines giving up smoking when the harm becomes tangible. Most of all we hope for a good life in the immediate future and would rather put aside unpleasant thoughts of doom to come.

We cannot regard the future of the civilized world in the same way as we see our personal futures. It is careless to be cavalier about our own death. It is reckless to think of civilization's end in the same way. Even if a tolerable future is probable it is still unwise to ignore the possibility of disaster.

One thing we can do to lessen the consequences of catastrophe is to write a guidebook for our survivors to help them rebuild civilization without repeating too many of our mistakes. I have long thought that a proper gift for our children and grandchildren is an accurate record of all we know about the present and past environment. Sandy and I enjoy walking on Dartmoor, much of which is featureless moorland. On such a landscape it is easy to get lost when it grows dark and the mists come down. We usually avoid this mishap by making sure that we always know where we are and what path we took. In some ways our journey into the future is like this. We can't see the way ahead or the pitfalls but it would help to know what the state is now and how we got here. It would help to have a guidebook written in clear and simple words that any intelligent person can understand.

No such book exists. For most of us, what we know of the Earth comes from books and television programmes that present either the single-minded view of a specialist or persuasion from a talented lobbyist. We live in adversarial, not thoughtful, times and tend to hear only the arguments of each of the special-interest groups. Even when they know that they are wrong they never admit it. They all fight for the interests of their group while claiming to speak for humankind. This is fine entertainment, but what use would their words be to the survivors of a future flood or famine? When they read them in a book drawn from the debris would they learn what went wrong and why? What help would they gain from the tract of a green lobbyist, the press release of a multinational power company, or the report of a governmental committee? To make things worse for our survivors, the objective view of science is nearly incomprehensible. Scientific papers and books are so arcane that scientists can only understand those of their own speciality. I doubt if there is anyone, apart from these specialists, who can understand more than a few of the papers published in *Science* or *Nature* every week.

Scan the shelves of a bookshop or a public library for a book that clearly explains the present condition and how it happened. You will not find it. The books that are there are about the evanescent things of today. Well-written, entertaining, or informative they may be, but almost all of them are in the current context. They take so much for granted and forget how hard won was the scientific knowledge that gave us the comfortable and safe life we enjoy. We are so ignorant of those individual acts of genius that established civilization that we now give equal place on our bookshelves to the extravagance of astrology, creationism and homeopathy. Books on these subjects at first entertained us or titillated our hypochondria. We now take them seriously and treat them as if they were reporting facts.

Imagine the survivors of a failed civilization. Imagine them trying to cope with a cholera epidemic using knowledge gathered from a tattered book on alternative medicine. Yet in the debris such a book would be more likely to have survived and be readable than a medical text.

What we need is a book of knowledge written so well as to constitute literature in its own right. Something for anyone interested in the state of the Earth and of us – a manual for living well and for survival. The quality of its writing must be such that it would serve for pleasure, for devotional reading, as a source of facts and even as a primary school text. It would range from simple things such as how to light a fire, to our place in the solar system and the universe. It would be a primer of philosophy and science – it would provide a top-down look at the Earth and us. It would explain the natural selection of all living things, and give the key facts of medicine, including the circulation of the blood, the role of the organs. The discovery that bacteria and viruses caused infectious diseases is relatively recent; imagine the consequences if such knowledge was lost. In its time the Bible set the constraints for behaviour and for health. We need a new book like the Bible that would serve in the same way but acknowledge science. It would explain properties like temperature, the meaning of their scales of measurement and how to measure them. It would list the periodic table of the elements. It would give an account of the air, the rocks, and the oceans. It would give schoolchildren of today a proper understanding of our civilization and of the planet it occupies. It would inform them at an age when their minds were most receptive and give them facts they would remember for a lifetime. It would also be the survival manual for our successors. A book that was readily available should disaster happen. It would help bring science back as part of our culture and be an inheritance. Whatever else may be wrong with science, it still provides the best explanation we have of the material world.

It is no use even thinking of presenting such a book using magnetic or optical media, or indeed any kind of medium that needs a computer and electricity to read it. Words stored in such a form are as fleeting as the chatter of the internet and would never survive a catastrophe. Not only is the storage media itself short-lived but its reading depends upon specific hardware and software. In this technology, rapid obsolescence is usual. Modern media is less reliable for long-term storage than is the spoken word. It needs the support of a high technology that we cannot take for granted. What we need is a book written on durable paper with long-lasting print. It must be clear, unbiased, accurate and up to date. Most of all, we need to accept and to believe in it at least as much as we did, and perhaps still do, the World Service of the BBC.

In the dark ages of our earlier history the religious orders in their monasteries carried through the essence of what makes us civilized.

Much of this knowledge was in books, and the monks took care of them and read them as part of their discipline. Sadly, we no longer have callings like this. The vast collection of knowledge that is now available is more than any one person could hold. Consequently it is divided and subdivided into subjects. Each subject is the province of professionally employed specialists. Most are expert in their own subject but ignorant of the others – few have a sense of vocation.

Apart from isolated institutes like the National Centre for Atmospheric Research perched on a mountain side in Colorado, there are no equivalents of the monasteries. So who would guard the book? A book of knowledge written with authority and as splendid a read as Tyndale's Bible might need no guardians. It would earn the respect needed to place it in every home, school, library and place of worship. It would then be to hand whatever happened.

Meanwhile in the hot arid world survivors gather for the journey to the new Arctic centres of civilization; I see them in the desert as the dawn breaks and the sun throws its piercing gaze across the horizon at the camp. The cool fresh night air lingers for a while and then, like smoke, dissipates as the heat takes charge. Their camel wakes, blinks and slowly rises on her haunches. The few remaining members of the tribe mount. She belches, and sets off on the long unbearably hot journey to the next oasis.

Glossary

algae

Algae are photosynthetic organisms that use sunlight to make organic matter and oxygen. The ocean plants are almost all algae; some are single cells, others, like kelp, can exist as huge assemblies of cells as long as sixty metres. The first algae on Earth appeared soon after life started over three billion years ago. Their form was bacterial and these microscopic organisms are still abundant: they are found either in living organisms or, importantly, as inclusions within the more complex cells of plants, called chloroplasts. Algae are unusually influential in the Earth's climate: they remove carbon dioxide from the air, and they are the source of the gas dimethyl sulphide (DMS) which oxidizes in the air to become the tiny nuclei that seed the droplets of clouds. Their growth in the surface waters of the sea is sensitively dependent upon its temperature, and if this is above 10 to 12°C the physical properties of the ocean prevent them from receiving nutrients and they do not flourish. Fossilized algae are the source of petroleum.

biosphere

The Swiss geographer Edward Suess coined the word 'biosphere' in 1875 for the geographical region of the Earth in which life is found. In this sense it is a precise and useful term and similar to the atmosphere and the hydrosphere, which respectively define where the air and the water is on Earth. In the second part of the twentieth century the Russian mineralogist V. Vernadsky expanded the definition of biosphere to include the concept that life was an active participant in geological evolution, and he encapsulated the notion in the phrase, 'Life is a geological force.' Vernadsky was following a tradition set by Darwin, Huxley, Lotka, Redfield and many others, but unlike them his ideas were mostly anecdotal. Biosphere is now mainly used, in Vernadsky's sense, as a vague, imprecise word that acknowledges the power of life on Earth without surrendering human sovereignty. Conveniently, it avoids any commitment to Gaia or Earth System Science.

chaos and chaos theory

Certainty and confidence in science marked its development in the nineteenth and much of the twentieth centuries, but now, like a battlefield hero fatally wounded, it carries on unaware that the determinism that had so long enlivened it was dead. The recognition that science was provisional and could never be certain was always there in the minds of good scientists, and the nineteenth-century application of statistics, first in commerce and then in science, made probabilistic thinking more intelligible than faith-based certainties. It took the discovery of the utter incomprehensibility of quantum phenomena to force the acceptance of a statistical more than a deterministic world; this was later consummated by the discoveries that came from the availability of affordable computers. These have enabled scientists to explore the world of dynamics, the mathematics of moving, flowing and living systems. The insights from the numerical analyses of fluid dynamics by Edward Lorenz and of population biology by Robert May revealed what is called 'deterministic chaos'. Systems like the weather, the motion of more than two astronomical bodies linked by gravitation, or more than two species in competition, are exceedingly sensitive to the initial conditions of their origin, and they evolve in a wholly unpredictable manner. The study of these systems is a rich and colourful new field of science enlivened by the visual brilliance of the strange images of fractal geometry. It is important to note that efficient dynamic mechanical systems, such as the autopilot of an aircraft, are essentially free of chaotic behaviour, and the same is true of healthy living organisms. Life can opportunistically employ chaos, but it is not a characteristic part of its normal function.

consilience

The evolutionary biologist E. O. Wilson, when writing on the incompatibility of twentieth-century science and religion, was mindful of the unconscious need in most of us for something transcendental, something more than could come from cold analysis. He disinterred the long-disused but still warm and worthy word, 'consilience', and offered it as something to link the thoughts of reductionist scientists with other intelligent humans, especially those with faith. I think he saw it as the name of a concept that would allow these two apparently irreconcilable concerns to evolve, if not together, at least in parallel. His thoughts are wonderfully well expressed in his book *Consilience* (1998).

Earth System Science
A discipline that has grown within the Earth science community among those dissatisfied with traditional geology as an intellectual environment for explaining the flood of new knowledge about the Earth. In particular, Earth System Scientists dislike the division of Earth and life sciences into the geosphere and the biosphere; they prefer to regard the Earth as a single dynamic entity within which the material and living parts are tightly coupled. This concept, together with its conclusion that the Earth self-regulates its climate and chemistry, was publicly stated in the Amsterdam Declaration of 2001. It differs from Gaia theory only because it has not had time to digest the mathematical consequences of the union between the Earth and life sciences, the most important of which is that self-regulation requires a goal. In Gaia theory the goal is to keep the Earth habitable for whatever are its inhabitants.

Gaia hypothesis
James Lovelock and Lynn Margulis postulated in the early 1970s that life on Earth actively keeps the surface conditions always favourable for whatever is the contemporary ensemble of organisms. When introduced it was contrary to the conventional wisdom that life adapted to planetary conditions as it and they evolved in their separate ways. We now know that both the hypothesis as originally stated and the conventional wisdom were wrong. The hypothesis evolved into what is now Gaia theory and the conventional wisdom into Earth System Science.

Gaia theory
A view of the Earth that sees it as a self-regulating system made up from the totality of organisms, the surface rocks, the ocean and the atmosphere tightly coupled as an evolving system. The theory sees this system as having a goal – the regulation of surface conditions so as always to be as favourable as possible for contemporary life. It is based on observations and theoretical models; it is fruitful and has made ten successful predictions.

greenhouse effect
Most of the sun's radiant energy is in the visible and near infra-red. The air, when free of clouds and dust, is as transparent to this radiation as is the glass of a greenhouse. Surfaces on the Earth, or within the greenhouse, are

warmed by sunlight, and some of this warmth is transferred to the air in contact with the surfaces. The warm air stays in the greenhouse mainly because the walls and glass roof prevent the restless wind from dissipating it. The Earth is kept warm in a similar but not identical way, by the absorption of radiant heat emitted from the warm surface by the gases carbon dioxide, water vapour and methane. These gases, although transparent to light, are partially opaque to the longer wavelengths emitted by a warm surface. This, the greenhouse effect, has long kept the surface air warm and, in the absence of pollution, is benign; without it the Earth would be 32°C colder and probably incompatible with life.

life

Because life exists simultaneously in the separated realms of physics, chemistry and biology, it has no decent definition. Physicists might define it as something that exists within bounds, that spontaneously reduces its entropy (disorder) while excreting disorder to the environment. Chemists would say that it is composed of macromolecules containing mainly the elements carbon, nitrogen, oxygen and hydrogen, and lesser but required proportions of sulphur, phosphorus and iron, together with a suite of trace elements including selenium, iodine, cobalt and others. Biochemists and physiologists would see life as always existing within cellular boundaries that hold an aqueous environment with a tightly regulated composition of ionic species including the elements sodium, potassium, calcium magnesium and chlorine; each of the cells carries a complete specification and instruction set written as a code on long linear molecules of deoxy ribonucleic acids (DNA). Biologists would define it as a dynamic state of matter that can replicate itself; the individual components will evolve by natural selection. Life can be observed, dissected and analysed but it is an emergent phenomenon and may never be capable of rational explanation.

natural ecosystems and ecosystem services

The phrase 'ecosystems services' was introduced by the biologist Paul Ehrlich and his colleagues in 1974 to acknowledge that an ecosystem was more then a place where biologists could study biodiversity, and that ecosystems were valuable as local regulators of climate, water and chemical resources. It is a useful term when used in this local sense about an ecosystem such as a tropical forest, but becomes vague, imprecise and too often anthropocentric when applied globally. Like 'biosphere' it is sometimes used as an escape from the more difficult concepts of Gaia or Earth System Science.

positive and negative feedback
When a car we are driving deviates from our intended path we alter the direction of the front wheels sufficiently to cancel the deviation. The error we have sensed is amplified by the power steering and applied to oppose the error. This is negative feedback. If by accident the steering mechanism was faulty and it increased, not opposed, the car's deviation, the error would be amplified and this would be an example of positive feedback. This is often a recipe for disaster, but positive feedback can be essential to making a system lively and rapidly responsive. When we talk of vicious circles we have positive feedback in mind, and this is the state the Earth appears to be in now; deviations of the climate are amplified not suppressed, so that greater heat leads to even greater heat.

renewable energy
The first law of thermodynamics states unequivocally that energy is always conserved, and therefore it can neither be lost nor renewed. When we talk colloquially of energy we are talking about the flow of energy, something that provides warmth, light, an ability to move, to communicate, and of course to sustain life. Renewable energy is a confusing concept that sounds good but defies close analysis. The energy taken by burning fossil fuels is said to be unrenewable, yet the carbon dioxide produced is used by plants, and ultimately a portion of it is buried to make more fossil fuel. The burning of crops grown as fuel is said to provide renewable energy, yet if we tried to fuel the world's present transport this way we would hasten, not delay, the onset of catastrophe. The land used to grow the fuel is needed for food and, more importantly, to sustain Gaia. With energy it is quantity not quality that matters. We can use any source we like as long as the total used is modest and does not hamper Gaia's economy.

rock weathering
Mountains continuously grow on the surface as the hot seething semi-fluid rocks beneath the surface drive the floating plates of rock into collision. On our timescale mountains are permanent features of the landscape, but in Gaian terms they are short lived and worn away by the weather. Rocks are cracked by frost, abraded by windblown sand and, most of all, dissolved away by rain. The dissolution of mountains by rain water is called by geochemists 'chemical rock weathering'; it happens because the rain contains dissolved carbon dioxide that reacts with the rocks to make water-soluble calcium bicarbonate. This solution is carried by the streams and rivers to the ocean. This fundamentally important sink for carbon dioxide was until

about 1980 considered by Earth scientists to be purely chemical. We now know that the presence of organisms from bacteria and algae on the rock faces and trees growing in the soil make a three to tenfold increase in rock weathering and carbon dioxide removal. It is fundamentally important for keeping the Earth cool and as part of Gaia's self-regulation.

Further Reading

CHAPTER 1 THE STATE OF THE EARTH

Stephen H. Schneider and Randi Londer, *The Coevolution of Climate and Life*, Sierra Club Books, San Francisco, 1984.
Stephen H. Schneider, *Global Warming*, Sierra Club Books, San Francisco, 1989.
John Gribbin, *Hothouse Earth and Gaia*, Bantam Press, London, 1989.
Stephen H. Schneider and Lynn Morton, *The Primordial Bond*, Plenum Press, New York, 1981.
John Gray, *Straw Dogs*, Granta, London, 2002.
John Gray, *Heretics*, Granta, London, 2004.
Ann Primavesi, *Gaia's Gift*, Routledge, London, 2003.
Fred Pearce, *Turning Up the Heat*, The Bodley Head, London, 1989.
Mary Midgley, *The Essential Mary Midgley*, Routledge, London, 2005.
Mary Midgley, *Science and Poetry*, Routledge, London, 2002.
Edward O. Wilson, *Consilience*, Little, Brown and Company, London, 1998.
Michael Crichton, *State of Fear*, Harper Collins, New York, 2004.
Michael Crichton, *Time Line*, Ballantine Books, New York, 1999.

CHAPTERS 2 AND 3 WHAT IS GAIA? THE LIFE HISTORY OF GAIA

John Gribbin, *Deep Simplicity*, Penguin, London, 2004.
Lynn Margulis, *The Symbiotic Planet*, Phoenix Press, London, 1998.
Lynn Margulis and Dorion Sagan, *Microcosmos*, Summit Books, New York, 1986.
Lee R. Kump, James F. Kasting and Robert G. Crane, *The Earth System*, Prentice Hall, New Jersey, 2004.

Richard Dawkins, *The Extended Phenotype*, W. H. Freeman, Oxford and San Francisco, 1982.
J. Scott Turner, *The Extended Organism*, Harvard University Press, Cambridge, Mass., 2000.
Edward O. Wilson, The *Diversity of Life*, Harvard University Press, Cambridge, Mass., 1992.
H.-J. Schellnhuber, *Earth System Analysis*, Springer, Berlin, 1998.
N. Morosovsky, *Rheostasis*, Oxford University Press, 1990.
Steven H. Strogatz, *Nonlinear Dynamics and Chaos*, Perseus Books, Cambridge, Mass., 2000.
Tim Lenton, 'Gaia and Natural Selection', *Nature*, 30 July 1998.
Tim Lenton and W. von Bloh, 'Biotic feedback extends Lifespan of Biosphere', *Geophysical Research Letters*, 28(a), 2001.

CHAPTER 4 FORECASTS FOR THE TWENTY-FIRST CENTURY

Sir John Houghton, *Global Warming*, Cambridge University Press, 2004.
Intergovernmental Panel on Climate Change, *Third Assessment Report, Climate Change 2001*, Cambridge University Press, 2001.
Millennium Ecosystem Assessment Report, Island Press, 2005.
Hubert Lamb, *Climate: Present, Past and Future*, Methuen, London, 1972.
Sir Crispin Tickell, *Climate Change and World Affairs*, Harvard University Press, Cambridge, Mass., 1986.
Kendall McGuffie and Ann Henderson-Sellers, *A Climate Modelling Primer*, Wiley, Chichester, 2005.

CHAPTER 5 SOURCES OF ENERGY

Rayner Joel, *Basic Engineering Thermodynamics*, Longman, Harlow, 1996.
W. J. Nuttall, *Nuclear Renaissance*, Institute of Physics Publishing, London, 2005.
Godfrey Boyle, *Renewable Energy*, Oxford University Press, 1966.
Fred Pearce, *Acid Rain*, Penguin, London, 1987.
Michael Laughton, *Power to the People*, ASI (Research) Ltd, London, 2003.
Neville Shute, *On the Beach*, Heinemann, London, 1961.
Helen Caldicott, *Nuclear Madness*, W. W. Norton, New York, 1994.
Bruno Comby, *Environmentalists for Nuclear Energy*, TNR, Paris, 2000.

CHAPTER 6 CHEMICALS, FOOD AND RAW MATERIALS

Bruce Ames, 'Dietary Carcinogens and Anticarcinogens', *Science*, Vol. 221, 1256–64, 1983.
H. A. Bridgman, *Global Air Pollution: Problems for the 1990s*, Belhaven Press, London, 1990.

CHAPTER 7 TECHNOLOGY FOR A SUSTAINABLE RETREAT

Robert A. Weinberg, *One Renegade Cell*, Basic Books, New York, 1988.

CHAPTER 8 A PERSONAL VIEW OF ENVIRONMENTALISM

Jonathon Porritt, *Seeing Green*, Blackwell, Oxford, 1984
Jonathon Porritt, *Playing Safe: Science and the Environment*, Thames and Hudson, London, 2000.
Rachel Carson, *Silent Spring*, Houghton Mifflin, Boston, 1962.
Richard Mabey, *Country Matters*, Pimlico, London, 2000.
Richard Mabey, *Nature Cure*, Chatto & Windus, London, 2005.
Edward Goldsmith, *The Way*, Shambhala, Boston, 1993.
Richard Rogers, *Cities for a Small Planet*, Faber & Faber, London, 1997.
Dick Taverne, *The March of Unreason*, Oxford University Press, 2005.

CHAPTER 9 BEYOND THE TERMINUS

Martin Rees, *Our Final Century*, Heinemann, London, 2003.

BOOKS ON GAIA

James Lovelock, *Gaia: A New Look at Life on Earth*, Oxford University Press, 1979.
James Lovelock, *The Ages of Gaia*, W. W. Norton, New York, 1988.

James Lovelock, *The Practical Science of Planetary Medicine*, 1991; reprinted as *Gaia: Medicine for an Ailing Planet*, Gaia Books, London, 2005.

James Lovelock, *Homage to Gaia*, Oxford University Press, 2000.

Index